THE BLACK BOX

THE BLACK BOX

ALL-NEW COCKPIT VOICE RECORDER ACCOUNTS OF IN-FLIGHT ACCIDENTS

Edited by Malcolm MacPherson

QUILL
WILLIAM MORROW
NEW YORK

It is the policy of William Morrow and Company, and its imprints
and affiliates, recognizing the importance of preserving what has been written,
to print the books we publish on acid-free paper,
and we exert our best efforts to that end.

Library of Congress Cataloging-in-Publication Data

The black box : all-new cockpit voice recorder accounts of in-flight accidents / Malcolm
MacPherson, editor.
p. cm.
ISBN 0-688-15892-7
1. Aircraft accidents. I. MacPherson, Malcolm.
TL553.5.B523 1998
363.12'4—dc21 98-9888
CIP

Printed in the United States of America

First Edition

20 19 18 17 16

BOOK DESIGN BY OKSANA KUSHNIR

www.williammorrow.com

ACKNOWLEDGMENTS

I would like to thank Ted Lopatkiewicz, deputy director of public affairs at the National Transportation Safety Board (NTSB) in Washington, D.C., for his help with these materials, and Bryan Reich, first officer, American Airlines, for vetting the manuscript with a professional's keen eye.

CONTENTS

INTRODUCTION

to the Revised Edition

Nearly everyone who has ever flown on a scheduled commercial flight has experienced the fear of flying. Whether in clear skies or cloudy, during a routine flight or in an emergency, we have shared a common feeling of dread, usually attended by clammy palms, whitened knuckles, and wildly racing hearts.

No education or sophistication, no reassuring baritone over the public-address system, no number of highballs, and no degree of embarrassment can seem to quell this fear when it comes to haunt us in those minutes between takeoff and landing. A gust of wind violently shakes the airframe in which we are held virtual prisoners, and no mantralike recitation of safety facts will quiet the terrified voice inside that is saying, "I *really* don't want to be here."

Still, the truism applies: Besides propulsion on our own two legs, the safest means of transportation is aboard a commercial airliner. As confusing as it may seem to our sense of logic, a commercial flight is safer even than eating, which should be food for thought the next time you're served an in-flight meal.

Thanks to the high level of professionalism of aircraft manufacturers, airlines (through their entire hierarchy), and especially of the airlines' flight crews, plus the vigilance of the Federal Aviation Administration (FAA) and the attention to detail of the National Transportation Safety Board (NTSB), aviation accidents have (and I apologize for this word) plummeted since the advent of big-jet travel in the early 1960s.

In America, about a hundred people die on average each year in commercial airline accidents. In 1996, a year of higher-than-average aviation death tolls, 1,187 people were killed on commercial jet flights *worldwide* in Western-built jets (394 in America, reflecting mostly two accidents—TWA 800 off Long Island, New York, with 230 fatalities and ValuJet in Florida with 110 fatalities). That 394 compares with 19,000 Americans that year who were murdered, 41,907 who succumbed in automobile accidents, and 714 who died in boating mishaps. In 1993, *one* person died in an aviation-related incident in America, and he was struck by a whirring propeller blade while standing on the ground.

In sum, for every million commercial airline hours flown, the fatal accident rate in the United States has fluctuated over the recent decades from a low of 0.013 to a high of 0.7; it was 0.439 in 1996, which was the fifth highest in the last fifteen years. In Europe the average was 0.9, and in Australia, 0.3. In regions of South America and Africa, China, and Russia, the rates rise. In Africa as a whole, there is an average of 13 fatal accidents per million hours; in South America, 5.7, according to the Flight Safety Foundation in Alexandria, Virginia.

In the years since publication of the first edition of *The Black Box* in 1984, the accident rate of airlines has leveled off. A reassuring explanation for this fact also reveals a disturbing new trend. Accidents are so rare in America and Europe that virtually anything the authorities and the airlines do to make flying safer will not reduce the accident rate by any appreciable statistical means. Yet as more and more people in parts of the world outside America and Europe begin to fly, according to a December 9, 1996, report in *The New York Times*, "the industry will have to lower its accident rate just to keep the number of crashes roughly constant each year." The report then reached an ominous conclusion: "If the current accident rate remains steady and air travel continues to grow rapidly, a passenger jet may crash as often as once a week by the year 2010."

Yet it is important to keep this in perspective. Flying in most "dangerous" parts of the world is still far, far safer than driving a car in the United States.

The commercial aviation industry is continually striving to make air travel safer. An enormous amount of work is constantly under way to find ways to improve safety. Wind shear *was* a big problem, but after several fatal wind-shear-related accidents (like Delta Flight 191 at Dallas–Fort Worth in August 1985), the industry sought to understand the phenomenon, how it forms, and what pilots should do about it. As a result, techniques to fly out of wind shear were developed. Now, with the help of research and new technologies, wind-shear incidents have decreased substantially. Add to that the introduction of passive safety measures like ground-proximity warning systems shouting at pilots to pull up, floor lighting in cabins to direct passengers during emergency exits, smoke retardants in manufactured materials, and aids like smoke detectors and fire extinguishers in bathrooms and in air-freight holds, and flying just keeps getting safer, in spite of burgeoning numbers of world flight departures and passengers.

Why then does the craven voice in our heads keep telling us when we are flying, "I *really* don't want to be here"?

For one thing, statistics are other people. For another, the media blow their coverage of air fatalities out of all proportion to their coverage of other forms of transportation deaths, and they frighten us. (When was the last time the report of an automobile accident, except for that of Princess Diana, appeared on the front page of *The New York Times*?) The media rationale is clear: When there is a commercial airplane accident, it often means a massive loss of human life. This also may provide the media with mysteries that resist a simple solution, like TWA Flight 800, the Boeing 747 that exploded after takeoff from JFK International Airport in 1996; the destruction of Flight 800's hull involved an unexplained, instantaneous horror. With the ValuJet crash in the Everglades on May 11, 1996, the hull was swallowed up in a swamp teeming with alligators, and until oxygen generators were discovered as the culprit, this crash also offered what looked like a very dark mystery. And the crash in Pittsburgh of USAir Flight 427 on September 8, 1994, still stubbornly defies a simple explanation for the loss of those 132 lives.

Even in the case of those that are easily explained, there is no denying that airline disasters hold a fascination for the media and the public alike that we can't really account for.

The airlines do all they can to reassure us with more than statistics and soothing words. They try to get us, as passengers, involved; they want us to *think* safety. There are obvious practical reasons behind seat belts, NO SMOKING signs, floor exit light strips, and straight seat backs during the critical flight periods of takeoff and landing; cabin crews do not demonstrate procedures with oxygen masks, flotation devices, and emergency exits without the certain awareness that these measures can prevent accidents and save lives. These measures also reassure us by telling us indirectly that the airline is concerned about our safety. Indeed, we may even think we hold a modicum of control over our destinies with our seat belts snugly around our laps. So it is also with the cockpit crews whose authoritative tones over the public-address system comfort us even when—especially when—they may be frenzied in an emergency.

Contributing to the mystery of our fears, while we may feel nothing out of the ordinary in the passenger cabin, a drama we never suspect may be unfolding in the cockpit. When things go wrong behind that closed door, often one crippled or failed mechanical system collapses on another. Suddenly, horns blare, recorded voices shout, "Terrain! Terrain! Pull Up! Pull Up!" Lights flash before the flight crew's eyes. Confusion often results; automatic responses are triggered in men and women who are more highly trained, retrained, and regularly scrutinized for performance than any other professionals in our society—including heart surgeons. In a truly amazing number of incidents we may never read about, these crews bring their airplanes back to land safely, without the passengers' having been aware that their lives were at risk.

In a tiny percentage of flights, airplanes crash, passengers die. The one true record of what went wrong is contained in recorders called black boxes, which federal authorities have required since 1957 as standard equipment aboard all commercial aircraft over a certain weight. Actually the boxes are painted a bright international orange with two white diagonal stripes for easy recognition and, stenciled in black on one side, the words FLIGHT RECORDER DO NOT OPEN. On the other side the words are in French: ENREGISTREUR DE VOL NE PAS OUVRIR.

The black box holds two components. One, the Flight Data Recorder (FDR), keeps a record of the airframe's performance—its compass headings, airspeeds, altitude levels, vertical acceleration, and microphone keying, among other things. The second component, required by federal law in commercial aircraft since 1966, is the Cockpit Voice Recorder (CVR), which weighs 21.5 pounds and records the sounds in the cockpit.

Within the orange outer CVR black box, a separate box of welded quarter-inch stainless steel protects (from up to 3,300 g's of force) a self-erasing, thirty-minute magnetic tape loop, which is being replaced in all new commercial aircraft by a two-hour self-erasing solid-state recorder. Black boxes are fitted in the tails of airframes; cables down the length of the fuselage connect them to one-inch circular microphones in the overhead area of the cockpit and the "boom mikes" (like the microphone that Madonna wears in her live performances) that FAA regulation requires pilots and copilots to wear while flying under 18,000 feet of altitude. Both black boxes (CVR, FDR) contain underwater locator beacons, which emit ultrasonic pings every second for sixty days after a crash, to help rescuers find them. The boxes are incredibly resilient. The force of impact has never destroyed one, even the recorder that sustained a crushing 9,000 g's of impact on board the hijacked PSA flight that crashed in 1987 near San Luis Obispo; fires rendered a few of the recorders useless before new federal fire regulations upgraded them a few years ago. And only one set of boxes is still lost on the ocean floor—that of an Alitalia jet thought to have been shot down off Libya in the Mediterranean by Muammar Qaddafi's air force in the early eighties.

I first started reading the transcripts of CVR tapes when I was living in Nairobi, Kenya, working as a correspondent for *Newsweek*. I never thought of myself as a ghoul for this unusual fascination. Far from it. I was able to picture the drama of what these transcripts contained, and they were unlike any I had ever read, seen on the screen, or watched in a theater. They were *real*, relived on the printed page.

Originally, I picked up the transcripts out of a concern about my own safety in the air. In those days I was taking off and landing in places and on airplanes that would naturally raise the fears in any sane person. (I once took off from Nouakchott, Mauritania, with live chickens in the cabin, and as we zoomed down the runway I spied a silver puddle on the taxiway out in the Saharan sun that I recognized as my Zero Haliburton suitcase.) In those days I believed that the more I learned about the behavior of crews in accident environments, the less I would fear flying.

Since then, I have tried to read the findings of the NTSB for most major commercial airplane accidents. Nothing I can think of has given me a better feel for what it's like up there when things go wrong. Unfortunately, while the transcripts continue to hold my interest for their drama, they have not lessened my fear of flying. (I read recently that the cockpit crew in a China Northern Boeing MD-82 over western China in November 1993 lost track of their altitude in the fog on a landing approach. The recorded ground-proximity warning system alerted them in English to "Pull up. . . . Pull up." The pilot was overheard later on the CVR, after they had crashed, asking his copilot in Chinese, "What means 'Pull up'?" I wondered then whether my fears were not well founded, even if I never plan to fly anywhere near China.)

No, the CVR transcripts have certainly not allayed my fears, nor probably will they yours. But often they can go a long way to explain in the most graphic and dramatic terms the professionalism, if not the outright heroism, of many commercial pilots and crews and the extraordinary measures they take to ensure our safety.

In the transcripts you are about to read, I have made no effort to "characterize" the crew members whose voices are taken directly off the CVR tapes. I do not want these transcripts to seem like an Arthur Hailey–type airport novel, as good as those may be. Whether the captain of the airplane is kind to animals or is married with children does not seem relevant to me in an accident environment; the same with the passengers whose lives are unknown to me. But I have tried to give the readers a context (of weather outside the airframe, of time, of the numbers of people in the cabin, the sights and the sounds). I have edited some of the dialogue for clarity and understanding, and I have put some of the pilots' jargon in brackets in layman's terms; a glossary serves as a further aid to understanding.

Readers of *The Black Box* might be advised to imagine themselves, rather than riding in these airplanes, tuned to an emergency frequency on a radio. You will be overhearing—albeit *reading*—the conversations between the cockpit crew members of an airplane in an emergency and, from time to time, conversations between the cockpit and the ground controllers giving assistance to the crippled flight. Even if you cannot visualize everything, I hope you will agree with my long-standing conviction that these transcripts are as dramatic reading as you are likely to find, because they are the minute-to-minute, unvarnished accounts of what actually happened.

—MALCOLM MACPHERSON

E-mail: m.macpherson@erols.com

THE BLACK BOX

East Grandby, Connecticut

November 12, 1995

■

American Airlines Flight 1572

I have chosen to lead the collection with this CVR tape because it shows how a cockpit crew, by their skill and resourcefulness, saved the passengers from what was otherwise certain disaster.

Aboard American Airlines Flight 1572 that blustery fall evening there were seventy-three passengers, including one child, with a flight crew of three cabin attendants, a pilot, and a copilot, who had left Chicago-O'Hare International Airport at 11:05 P.M. after a two-hour delay. The flight to Bradley International Airport (BDL) at Hartford, Connecticut, was largely uneventful, although the captain changed from a flight altitude of 33,000 feet to 35,000 feet to avoid turbulence. For the most part, the passengers rode comfortably.

Near the flight's conclusion, about three hundred miles from the airport, Flight 1572 was cleared direct to Bradley. Outside the aircraft, a low-pressure area that extended south from Quebec over New York City caused strong southerly and westerly winds and rain. At times at the airport toward which Flight 1572 was heading, the winds gusted to forty-nine miles an hour. As the flight approached BDL, the crew received two messages: one notifying them of a changed altimeter setting and another that warned of turbulence and wind shear on final approach.

At thirty-two minutes after midnight, the flight descended to 19,000 feet. Severe turbulence was reported below 10,000 feet. At 12:32:23, the captain advised the flight crew to secure the cabin because of the anticipated turbulence. The crew was briefed by ground controllers to approach Runway 15. At 12:45, the crew was further advised by Bradley's Approach Control to descend to 4,000 feet. Four minutes later, the flight crew was told that the Tower was closing temporarily to repair a window loosened by the high winds.

On its landing approach, the aircraft went below the minimum descent altitude of 1,008 feet. The pilot glanced out the window. When he looked back at his instruments, the airplane was flying at only 350 feet—five seconds from hitting a cluster of tall conifers near the end of Runway 15. An outdated altimeter setting had allowed the autopilot to descend the aircraft about 309 feet below the indicated minimum descent altitude. The leading and trailing edges of the wings sheared off the tops of the trees. Power from one engine failed almost instantly; power from the other failed shortly thereafter.

We pick up the CVR tape as the flight is just starting its descent into Bradley.

FLIGHT ATTENDANT: [*To captain on interphone*] Were you looking for me?

CAPTAIN: We just got the message. It's . . . going to be real bumpy on the way down to landing.

FLIGHT ATTENDANT: Okay.

CAPTAIN: So put everything away or whatever as soon as we start down. Uh, matter of fact, we're starting our descent now. So you can lock it all up and prepare for landing.

FLIGHT ATTENDANT: Okay, thank you.

CAPTAIN: [*To copilot*] Let's go down.

COPILOT: Out of three five oh [35,000 feet] for flight level one nine oh [19,000], American Fifteen seventy-two.

CAPTAIN: [*To copilot*] You might want to cool it down, too. It's gonna get bumpy [and the passengers will] be throwing up.

Bradley Tower advises American 1572 that there is now severe turbulence below 10,000 feet.

BRADLEY: Bradley Airport information Victor, zero three five one . . . temperature six two, dew point five seven, wind one six zero at two eight, gust three niner, altimeter two niner five zero. Approach in use, ILS [Instrument Landings System] Runway Twenty-four or VOR [Very high frequency Omnidirectional Radio range, which is an instrument landing directional beacon] Runway Fifteen.

A minute later, the copilot calls Bradley for another report on the weather at the airport. The information about wind shear and severe turbulence is reported.

CAPTAIN: [*To passengers on public-address system*] We started our descent. Now we're about a hundred miles away from Bradley Field right now. Be touching down in about twenty-five minutes or so. . . . And the latest temperature's sixty-two degrees and just calling it, uh, light rain. However, the winds [are] pretty, uh, pretty high. They're saying the winds are up to thirty miles an hour or so. So it might get a little choppy right now. They're reporting some moderate turbulence on the descent. Might just get a little choppy on the way down.

CAPTAIN: [*To copilot*] Just watch me the whole way, all right?

COPILOT: Yeah, man. You got it.

CAPTAIN: Any comments, scream out.

COPILOT: You're going to get a lot of turbulence. You know how to land it. I'll tell you, when I was an engineer, new with the airlines, I watched a guy. He's, uh—nobody likes him. He's a [Boeing] Seven two [seven] captain. I forgot his damn name. All the [former] navy guys, man, they hate his guts. He's still on the Seven two [seven]. He thinks he's an IP [instructor pilot] at the RAG [training squadron].

CAPTAIN: Uh.

COPILOT: I watched him land in thirty knots of direct cross[wind] up at Bradley when I was a wrench [flight engineer], and it scared the hell out of me. Actually it scared me bad. Some of the flying we do here [at Bradley] is much harder than . . .

CAPTAIN: Yeah, yeah, I agree, I agree.

COPILOT: Just fightin' it all the way down.

CAPTAIN: Oh, what gate are we goin' into?

COPILOT: Gate Eight.

CAPTAIN: Gettin' a lotta rain out there.

COPILOT: Altimeters?

CAPTAIN: I'll tell you, flying at night. I don't like it worth a damn. [Altimeters at] twenty-nine fifty?

COPILOT: Yeah, they called twenty-nine forty-seven when we started down.

CAPTAIN: Okay.

COPILOT: Pumps are up. You want these lights?

CAPTAIN: I think we can leave them [lights] off for now.

CAPTAIN: [*Gets on public-address system*] Flight attendants prepare for landing, please.

COCKPIT: [*Yawns*]

CAPTAIN: [*To copilot*] What's the overcast?

COPILOT: Twenty-seven hundred [feet].

BRADLEY APPROACH CONTROL: American Fifteen seventy-two, fly heading of, uh, one zero zero.

COPILOT: One zero zero, American Fifteen seventy-two.

CAPTAIN: One zero zero, one oh nine, one forty-eight. I'll use medium brakes.

COPILOT: Can't see [the airport] yet.

CAPTAIN: Lotta rain.

COPILOT: I can see that.

APPROACH: American Fifteen seventy-two, descend and maintain four thousand [feet].

COPILOT: Eleven for four thousand, American Fifteen seventy-two.

APPROACH: American Fifteen seventy-two, the winds are one seven zero at two nine, gust three nine.

COPILOT: Copy.

COCKPIT: [*Laughter*]

CAPTAIN: How about slats [forward edge flaps on wings]. Extend [them], please.

COCKPIT: [*Sound of rattling, like aircraft going through turbulence*]

COPILOT: Approaching four thousand [feet of altitude].

CAPTAIN: Okay. Slowing down.

APPROACH: American Fifteen seventy-two, you're five miles from [final approach] at or above three thousand five hundred feet, cleared for VOR Runway Fifteen approach.

COPILOT: Okay, we're cleared for the approach, and we'll cross the [final] at or above, uh, thirty-five hundred. American Fifteen seventy-two.

CAPTAIN: Set. Comin' down.

APPROACH: American Fifteen seventy-two, be advised the Tower is closed at this time. It's a temporary closure, due to a problem with one of the windows, so I'll need a down time on you, but you can stay on this frequency for that.

COPILOT: Roger. What happened on the window?

APPROACH: It's just loose. They've got carpenters up there now boarding it up. . . . But once that's done the Tower should be open.

COPILOT: Copy.

CAPTAIN: Flaps eleven, please.

COPILOT: You got it.

COCKPIT: [*Sound of aircraft going through turbulence*]

COPILOT: Okay, thirty-five hundred feet. Looking good. . . . I think it's gonna be smoother once we get out of the weather.

CAPTAIN: Yeah.

COPILOT: Okay, you're at thirty-five hundred.

CAPTAIN: Okay, we're cleared down to where?

COPILOT: You're cleared down to two thousand. . . .

CAPTAIN: Okay, two thousand set and armed.

COPILOT: Five miles, so it's good. Two thousand is set and armed.

CAPTAIN: Flaps fifteen [lower flaps].

COPILOT: Down to flaps fifteen.

COCKPIT: [*Sound of flap/slat handle being moved and sound of rattling, like aircraft going through turbulence*]

CAPTAIN: Okay, comin' down.

COPILOT: Ten miles.

CAPTAIN: Comin' back to idle.

COPILOT: Roger.

APPROACH: American Fifteen seventy-two, you show yourself on the final? Looks like you're, uh, a bit to the left of it.

COPILOT: Copy.

CAPTAIN: Yeah, looks like we're to the left of it.

APPROACH: American Fifteen seventy-two, roger, and the wind's now one seven zero at two four, gust three five [35 mph].

COPILOT: Roger.

CAPTAIN: How about [landing] gear down, please.

COCKPIT: [*Sound of landing-gear handle being operated, followed by sound of nose-gear door opening*]

COPILOT: . . . Thousand and five.

APPROACH: American Fifteen seventy-two, there is someone [back] in the Tower. It's not really officially open, but you can change to Tower frequency, one two zero point three.

COPILOT: Okay, you're not gonna need that down time [in the Tower]?

APPROACH: Negative.

COPILOT: See ya.

APPROACH: Good day.

CAPTAIN: Okay, it's two thousand feet until five miles [from runway].

COPILOT: That's it.

CAPTAIN: Comin' back. Flaps twenty-eight.

COCKPIT: [*Sound of flaps handle being moved*]

COPILOT: Okay, going down to nine oh eight [908 feet], huh?

CAPTAIN: Yeah.

COPILOT: Set and armed. VOR right on track.

CAPTAIN: Okay.

COPILOT: Gear's down and green spoiler lever?

COCKPIT: [*Sound of click, like spoiler lever being armed*]

CAPTAIN: Armed.

COPILOT: You got brakes. . . . [Brakes] are going to be medium.

COPILOT: [*To Approach*] Hey, Tower, American Fifteen seventy-two, we're on a six-miles final for Runway Fifteen.

APPROACH: American Fifteen seventy-two, landing is at your discretion, sir. The wind at one seven zero at two five, peak gust to four zero . . . and, uh, the runway does appear clear. You can land and taxi to the gate at your discretion.

COPILOT: [*To captain*] Showing you going through the course. [*To Approach*] What are you showing right now for winds?

APPROACH: One seven zero at two four.

COPILOT: Copy.

COCKPIT: [*Sound of rattling, like aircraft going through turbulence*]

CAPTAIN: Flaps forty.

COCKPIT: [*Sound of click, like flap/slat handle being moved*]

COPILOT: Annunciator lights checked, flaps and slats at forty-forty and land. You're cleared to land, dude.

CAPTAIN: Okay, give me a thousand down.

COPILOT: One thousand down. You got it.

APPROACH: Wind-shear alert, uh, center field one seven zero at two five. The northeast boundary, one seven zero at two four, one niner zero at twelve at the southeast boundary.

COPILOT: Copy.

COPILOT: [*To captain*] There's a thousand feet. You got forty-forty land, cleared to land.

CAPTAIN: Okay.

COPILOT: Nine hundred and eight feet is your, uh . . .

CAPTAIN: Right.

COPILOT: . . . Your bug [limit].

COCKPIT: [*Sound of turbulence*]

COPILOT: You're going below your [limit].

COCKPIT: [*Recorded mechanical voice: "Sink rate, sink rate."*]

COCKPIT: [*Sound of first impact*]

At little more than 900 feet above the ground and 2.54 miles from the approach end of Runway 15, American 1572 clips the tops of the trees, swerves sharply left, then sharply right, then left again in line with the runway.

COCKPIT: [*Recorded mechanical voice: "Sink rate." Sound of four beeps. "Wind shear, wind shear, wind shear."*]

COPILOT: Go, go around.

At this point, the captain pushes the throttle full open to abort the landing. Landing gear is put up and flaps are brought back to 15 degrees from 40.

CAPTAIN: We're going, going, going around, going around. . . .

COCKPIT: [*Recorded mechanical voice: "Landing gear." Sound of horn. Sound of four beeps. "Wind shear, wind shear, wind shear."*]

CAPTAIN: Flaps fifteen, positive rate [they are climbing], gear up.

COCKPIT: [*Recorded mechanical voice: "Landing gear."*]

COPILOT: You want the gear up?

CAPTAIN: Yep.

COCKPIT: [*Sound of horn. Recorded mechanical voice: "Landing gear, landing gear."*]

A couple of seconds after the airplane's wings shear off the treetops, the airspeed starts to decrease because of the damage to the two engines, and the airplane begins a slow descent. The rain has stopped. The copilot sees the runway ahead.

CAPTAIN: Left motor's failed.
COPILOT: There's the runway straight ahead.
CAPTAIN: Okay. Tell 'em [Bradley Tower] we're goin' down. Tell 'em emergency.
COPILOT: [*To Approach*] Tower, call for emergency equipment. We're going down on the runway. [*To captain*] You want the [landing] gear back down?
CAPTAIN: Yes, throw it down!
COCKPIT: [*Recorded mechanical voice: "Sink rate, sink rate."*]
CAPTAIN: Oh, god.
COPILOT: You're gonna make it.
CAPTAIN: Okay.
COCKPIT: [*Recorded mechanical voice: "Sink rate, sink rate."*]
COPILOT: Flaps?
CAPTAIN: Put 'em down. What we got?
COPILOT: We're still flying.
CAPTAIN: Okay. God . . .
COPILOT: Keep coming.
COCKPIT: [*Recorded mechanical voice: "Sink rate, too low, flaps, terrain, terrain, terrain, too low."*]
COPILOT: [*To Approach*] Call for emergency, call for emergency equipment.
APPROACH: They're coming, they're comin'.
COCKPIT: [*Recorded mechanical voice: "Terrain, too low."*]
COPILOT: You got it, dude. You're gonna make it.
CAPTAIN: Okay. Flaps forty. All the way down.
COCKPIT: [*Recorded mechanical voice: "Don't sink."*]
CAPTAIN: All the way, flaps forty.
COPILOT: They're all the way.
CAPTAIN: Okay, hold on, guy.
COCKPIT: [*Sound of impact, this time with the ILS antenna sticking up near the end of the runway*]
CAPTAIN: Hold it down, buddy, hold it down, hold it down, hold it down, hold it down. . . .

Flight 1572 slams down on the runway.

COPILOT: God bless you. You made it.
COCKPIT: [*Recorded mechanical voice: "Landing gear."*]

The airplane rolls to a stop on the runway.

COCKPIT: [*Sound of engine rpm decreasing*]
CAPTAIN: Shut down the motors.
COPILOT: Pull 'em both?
CAPTAIN: Yeah, pull both the fire handles.

END OF TAPE

The passengers and the crew members evacuated the airplane via the inflated door slides. There were no fatalities, and only one minor injury to the passengers. Damage to the airplane was estimated at $9 million.

Lima, Peru

October 2, 1996

■

AeroPeru Flight 603

This is one of the more wrenching CVRs in this collection. It chronicles the fate of AeroPeru Flight 603 from Lima, Peru, to Santiago, Chile.

Earlier that day, the ground crew at Lima International Airport, while washing the fuselage of the AeroPeru Boeing 757, had taped over the aircraft's left-side static ports with masking tape. These ports, the small metal "probes" that stick out of the fuselage of airplanes near the nose, read the ambient temperature, air pressure, and measured wind speed. With tape covering the openings, the static ports could not function. For the sophisticated Boeing 757, it was a little like being blind.

The ground crew had failed to remove the tape from the ports when they finished washing the aircraft. And during their preflight visual check ("walk-around") of the aircraft, the cockpit crew of Flight 603 failed to take notice of the tape.

As a result, the flight took off that night without instruments to read air speed or altitude. While the crew had the ability to control the aircraft, the autopilot and auto-throttle computer controls lacked the data to operate with any accuracy.

From the start, the crew struggled to make sense of the perilous situation for twenty-eight minutes, as the aircraft veered over the Pacific, changing course and altitude several times.

The sixty-one passengers and seven cabin attendants were never told what was wrong.

We pick up the CVR as Flight 603 is cleared for takeoff.

Copilot: [*To Tower*] Lima Tower, AeroPeru Six zero three, Runway Fifteen, ready for takeoff.
Lima Tower: AeroPeru Six zero three, use noise attenuation, wind calm. You are authorized to take off Runway Fifteen.
Pilot: One five, one five, transponder.
Copilot: Flaps one five, takeoff briefing complete. Time check.
Pilot: Let's go.
Cabin: [*Sound of engines accelerating*]
Pilot: Power's set.
Copilot: Power's set.
Pilot: Eighty knots.
Copilot: Checked.
Pilot: Vee one.
Pilot: Rotate.
Pilot: Vee two.
Pilot: Positive [climb rate].
Copilot: Gear up.
Cabin: [*Sound of landing-gear control moving*]
Pilot: Right, Vee two plus ten.
Copilot: The altimeters are stuck!

This is the crew's first observation that their instruments are inoperative.

Cabin: [*Sound of wind-shear alarm*]
Copilot: Hey, altimeters have stuck!
Pilot: Yeah.
Copilot: All of them?
Pilot: This is really new. . . . Keep Vee two plus ten. . . . Vee two plus ten.
Copilot: The speed . . .
Pilot: Eh?
Copilot: The speed . . .
Pilot: Yeah, right, Vee two plus ten . . .
Pilot: What happened, [we're] not climbing?
Copilot: I'm climbing, but the speed?
Pilot: Hold it, no. . . . Keep the speed.
Pilot: Positive, eh, put ten degrees. We are descending.
Copilot: Rudder ratio.

At this point the rudder-ratio alarm sounds, indicating abnormal operation of speed sensors. This warns the pilot to avoid large or abrupt rudder inputs. This is one of the signals that will hold the attention of the pilots throughout the emergency.

PILOT: How strange . . . Turn to the right. . . . Well . . . Go up! Go up! Go up! Go up! Go up! Go up!
COPILOT: I'm . . .
PILOT: Go up! You are going down. . . .
COPILOT: I am up, but the speed . . .
PILOT: Yeah, but it's [speedometer is] stuck. . . . Mach trim, rudder ratio . . . Now you are . . . Go up! Go up! Go up! Go up! Go up heading one hundred. . . . Eh . . . it's okay, this course.
COPILOT: Put the . . . Climb thrust . . . [Is the] center autopilot in command?
PILOT: I don't think so. . . . Let me see. . . .
COPILOT: No. The speed is better now.
PILOT: Yeah, shit. Rudder ratio.
COPILOT: Yeah. I have . . . let's see, source selector.

The pilot now seeks an alternate static source of input for the cockpit instruments.

PILOT: Yeah, shit. Rudder ratio.
COPILOT: Mach trim, Mach trim.
PILOT: The speed . . . Let's go to basic instruments. Everything has gone shit.
CABIN: [*Sound of caution alarm; sound of alert alarm*]
COPILOT: [*To Tower*] Tower, AeroPeru Six zero three.
PILOT: We are in emergency.
TOWER: AeroPeru Six zero three, Tower, go on.
COPILOT: [*To Tower*] Okay. We declare emergency! We have no basic instruments, no altimeter, no speedometer. We declare an emergency!
TOWER: Roger, altitude?
PILOT: Ahh . . .
COPILOT: [*To Tower*] We don't have, we have until thousand feet. . . . Approximately seventeen hundred.
PILOT: One thousand seven hundred.

TOWER: One thousand feet, roger, roger. AeroPeru Six zero three, confirm if you can change frequency one one niner point seven to receive instructions from Radar Control.
COPILOT: We go to one one nine point seven.
PILOT: Auto-throttle has disconnected.
COPILOT: Auto-throttle has disconnected.
PILOT: Let's see, what reads there? Let . . .

Pilot asks his copilot to look in the flight manual for information to help them out of their situation. Copilot reads the pertinent data.

COPILOT: Auto-throttle disconnect, rudder ratio, and Mach speed indicator.
PILOT: Okay.
COPILOT: Five hundred feet, it's grabbing now. These assholes from maintenance move everything.
PILOT: What shit have they done?
COPILOT: Auto-throttle, rudder ratio, Mach speed trim.
COPILOT: [*To Air-Traffic Control*] Let's see if you have us on the radar, Tower, AeroPeru Six zero three.
PILOT: I have the command, eh. Don't . . .

The pilot now takes control of the aircraft from the copilot.

LIMA AIR-TRAFFIC CONTROL [ATC]: AeroPeru Six zero three, Lima.
PILOT: Autopilots have been connected.

The pilot sees that the autopilot light is on.

COPILOT: No, no, they are disconnected.
PILOT: Ah, are they on?
COPILOT: Yes, but they are off, only flight director is on.
COPILOT: [*To ATC*] Lima, Six zero three.
PILOT: [*To copilot*] It's maintaining five degrees.
COPILOT: [*To ATC*] We request vectors for ILS, Runway Fifteen.

Copilot asks for a heading to return to Lima International. Flight 603 starts to turn out to sea, then north along the coast.

PILOT: Not yet, not yet, let's stabilize.
ATC: Correct, we suggest a right turn heading three three zero.
COPILOT: [*To ATC*] Turning right, course three three zero.

ATC: Affirmative, and keep present altitude.
PILOT: Altitude? We can't. . . . We are going up.
COPILOT: [*To ATC*] What level [of altitude] do we have? Do we have four thousand feet? Let's see if you confirm us.
ATC: Correct, keeping four thousand.
COPILOT: [*To ATC*] We are keeping four thousand.
ATC: Affirmative.
PILOT: Attitude, just attitude. The rudder ratio must be . . . Autothrottle disconnect.
COPILOT: Really . . . we don't have any control.
PILOT: We don't have control, not even the basics.
PILOT: Let's see, check everything.
COPILOT: Four thousand feet.
PILOT: Okay, never mind. We keep going up. We are flying without speed. It can't be. Let's see. . . . Respond.
COPILOT: Speed is zero, all speeds.
COPILOT: [*To ATC*] We, ah . . . The speed, please, if you have us on the radar?
ATC: Yes, affirmative, stand by ten seconds. It seems that you are going up with level six zero, at twenty-two miles south, and heading one niner five.
COPILOT: [*To ATC*] Okay. That is right. We have course one nine zero and we have seven thousand feet on the altimeter.
ATC: Yes, correct, now reaching seven zero.
COPILOT: [*To ATC*] We have control problems.

But Flight 603 does not have control problems. It has instrument problems and problems with indicators like rudder ratio and speed-trim signals.

PILOT: Rudder ratio.
ATC: Okay, roger.
PILOT: Let's see, read rudder ratio and Mach trim.
COPILOT: Flight controls . . . rudder ratio . . .
PILOT: Rudder ratio on or not?
COPILOT: Yes.
PILOT: Eh . . . yes.
COPILOT: Yes, they are on.
PILOT: Both.
COPILOT: "Avoid large or abrupt—"

Copilot now starts reading from the flight manual.

PILOT: What?
COPILOT: "Avoid large or abrupt rudder inputs, if normal left hydraulic system pressure available." Left hydraulic system available. Yes, crosswind limit. Do not attempt an auto-land, [the manual] says.
PILOT: Yeah, bullshit, we can't even fly!
COPILOT: I'm trying to set speed.
PILOT: But we don't even have . . .
COPILOT: Air data.
PILOT: Air data. Okay?
COPILOT: Alternate air data. I am setting it.
PILOT: Let's see, okay.
COPILOT: Nothing. Eight thousand . . . nine thousand . . .
COPILOT: They haven't given us anything.
PILOT: Here it is.
ATC: AeroPeru Six zero three, now you show level nine two hundred. . . . Ah? What is your course now?

Flight 603 is now over the Pacific, about to head north along the coast.

PILOT: We are flying two hundred.
COPILOT: [*To ATC*] Course two zero five.
ATC: Affirmative, you are turning slowly to the right, correct?
COPILOT: [*To ATC*] No, we are keeping course to fly away from the shore.
PILOT: No, we are trying to maintain course. We will try to keep ten thousand feet. Altitude hold, set it . . . ten thousand feet.
COPILOT: Fast is out now. Fast. Altitude hold.
ATC: Your distance is thirty miles. You want course for proceed to localizer, right?
COPILOT: [*To ATC*] Correct.
PILOT: Not yet. Let's solve the problem.
ATC: Turn right, heading three five zero, we suggest.
COPILOT: But we are going, ah . . .
PILOT: Take the autopilots out.
COPILOT: Autopilots are out.
PILOT: Yeah, I know.
COPILOT: It is leveled now.

PILOT: No. Go on.
COPILOT: We will set ten thousand feet here in the auto.
PILOT: Vertical speed . . . No, no, don't set vert speed hold, hold, altitude hold.

Here the cockpit crew tries to key new inputs into the command computer.

COPILOT: In eleven.
PILOT: Okay, in eleven thousand feet.
COPILOT: Twelve, let's put twelve then.
PILOT: Okay. Now it's indicating the speed.
COPILOT: [*To ATC*] Okay. We are recovering speed and we request keeping twelve thousand.
PILOT: Flaps . . . eh . . . five.

The pilot realizes that the flaps are still set at five degrees from takeoff configuration.

CABIN: [*Sound of flap commands moving*]
ATC: Correct, maintain level one two zero. Report starting turn. . . . Course to the right. Suggested course three five zero.
COPILOT: [*To ATC*] Roger.
PILOT: Decrease power to me. Flap up.
COPILOT: Flap up.
CABIN: [*Sound of flap commands moving*]
PILOT: Two twenty, we will keep two twenty [mph].
COPILOT: Twelve thousand feet.
PILOT: Go on reading [the manual].
COPILOT: Nothing else. The only thing I made is the air data comp.
PILOT: The auto-throttle . . . The autopilots are on.
COPILOT: They are not set on.

For the next several seconds the crew experiments with throttle and flap settings.

CABIN: [*Sound of alert*]
COPILOT: Now we are okay.
PILOT: Yes, now we are okay.
COPILOT: Do you want center autopilot?
PILOT: Yeah.

Connecting the autopilot was probably a mistake. The pilot should have continued to fly the airplane himself.

COPILOT: Let's put this.
PILOT: Take the speed out for me.
COPILOT: Center autopilot.
COPILOT: Pull here.
PILOT: Keep trying.
COPILOT: There is no auto-throttle.
PILOT: The speed . . . Autopilot off.
COPILOT: Airspeed . . . Autopilot off.
PILOT: At two twenty we will keep manual [controls].
COPILOT: Okay.
ATC: AeroPeru Six zero three, you are forty miles from Lima and according with information on screen at level one two zero [12,000 feet]. Approximate speed over the ground is three hundred and ten knots [357 mph].
PILOT: Roger.
COPILOT: [*To ATC*] Roger. We have two thirty and are turning to course three thirty.
ATC: Okay. Roger.
COPILOT: Rudder ratio follows. . . . Mach speed trim.
PILOT: Okay. Let's see, ah. . . . Read all that . . . auto-throttle is connect. Command, one two zero feet here, correct?
COPILOT: Twelve thousand feet.

Again, the crew relies on the autopilot.

COPILOT: Ah, auto-flight, now what? Auto-throttle, rudder ratio, I read it already. Mach speed trim, Mach speed is in navigation, flight instruments, flight instruments, auto-flight, auto-flight, altitude alert, autopilot disconnect, rudder ratio, okay.
PILOT: Hey, shit, we are going down now! Hey, shit, this auto-throttle . . . The autopilot is, fuck, off!
COPILOT: Speed, altitude alert, autopilot disconnect.
PILOT: Three thirty.
COPILOT: Autopilot inoperative, flight instruments switch, Mach speed, I don't find that Mach speed trim. Speed, we have speed problems, instrument source selector, flight director.

PILOT: It can't be. Hey, the speed. Look. The power we have. . . . It can't be!
COPILOT: It can't be, it's true, it's wrong. . . . Three thirty [knots].
PILOT: Yes, but they are even, aren't they?
COPILOT: Okay, set yours on alternate air data, the one down there, the lower button.
COPILOT: Oh, shit. Worse. Your altimeter goes to shit!
PILOT: Fuck. Basic instruments, let's go to basic instruments!
COPILOT: Basic instruments.
PILOT: Eh . . . the air data gone to shit there.
COPILOT: I'll fill here the departure and arrival. . . . ILS Runway Fifteen . . . Insert.

Now the copilot keys into the computer data needed for landing.

PILOT: Okay. Yours is screwed up. The air data.
COPILOT: Two four eight, two four eight, nineteen seven.
PILOT: We are with three three zero [heading].
COPILOT: With three thirty.
ATC: AeroPeru Six zero three, you are at forty miles, correct? Flying parallel course three three zero. You are flying left [of] course about to intersect west of Lima.
COPILOT: It is falling. . . .
COPILOT: [*To ATC*] Correct, we request vectors from this moment on.
PILOT: We are going to stabilize and . . . Let's see. . . .
ATC: Roger, we suggest course three six zero.
COPILOT: [*To ATC*] Three six zero.
ATC: The Lima VOR, do you receive it okay?
COPILOT: [*To ATC*] Affirmative.
PILOT: There is no flight director [autopilot]. No auto-throttle either. And where . . . where . . . ? Is this going up?
CABIN: [*Sound of engines accelerating*]
COPILOT: [*To ATC*] Okay. I understand, for . . . three one five, turn right to intercept localizer. Ah, we still don't have . . . ah . . .
PILOT: Auto-throttle off. Hey, you know, it's . . .
COPILOT: It's falling!
PILOT: Shit, yeah!
COPILOT: [Airspeed is] going up too much.

PILOT: It doesn't dis . . . apparently doesn't disconnect the auto—

CABIN: [*Sound of alert*]

COPILOT: Autopilot disconnected, better if we control it on basics, isn't it?

PILOT: Let's see the trim.

ATC: According to presentation [on radar], you are crossing radial two three zero from Lima, distance west southwest is thirty-seven miles.

COPILOT: [*To ATC*] Correct. We will . . . We have problems here for reading instruments. You will have to help us in altitudes and speed if possible.

ATC: Okay, roger.

COPILOT: [*To ATC*] Until we reach the localizer beam and we will navigate some thirty miles north of the localizer of Lima VOR to center in the ILS.

PILOT: It doesn't have alt[itude] hold. . . . Set the approach, please.

COPILOT: I did it, I did it.

PILOT: Then let's go.

COPILOT: Okay the approach . . .

PILOT: Let's see, we will try to go down. Okay with this course and . . . Why it doesn't appear . . . ? Ah . . . okay, okay. I'll try to go down with everything cut. . . . Ah . . . okay.

COPILOT: [The speed is] going up.

PILOT: Let's go down to ten thousand feet. . . . Why does speed get so high? Is it the real speed?

COPILOT: This is what worries me. . . . No, I don't think so. . . .

PILOT: It can't be. . . . Then . . . rudder ratio, Mach trim . . . Then if there is not that . . . Mach-speed trim?

COPILOT: No, there is no Mach-speed trim. I don't find . . . Let's see, keep flying it. It will . . . [*To ATC*] Can you tell us the speed, please?

ATC: It indicates three two zero [320 knots].

COPILOT: [*To ATC*] Correct, we have all engines cut and it's accelerating . . . accelerating.

COPILOT: Take the speed brake out.

PILOT: No . . . eh, no. Let's see, maybe. Let's see, take out . . .

COPILOT: Speed is okay. All three indicators are okay in speed. Fast, fast. All engine instruments are okay, all engine instruments are okay. Shit!

PILOT: What would be the real speed?

COPILOT: This one is okay. They are okay, the speed. . . . Airspeed . . .
PILOT: But with all power cut down, it can't be, with all cut down. . . . There's a problem with the source instrument. Let's see, how many miles . . . ? At thirty miles from Lima, we start descending with spoilers and flight-level change.
ATC: You are crossing the two six zero of Lima, at thirty-one miles west. Flight level is one hundred plus seven hundred, and approximate speed is two eight zero over the ground.
PILOT: Yeah . . . perfect.
COPILOT: [*To ATC*] Yeah, but we have an indication of three hundred fifty knots here.
PILOT: Let's put. Why are these more clear?
CABIN: [*A two-tone overspeed alarm sounds*]
PILOT: Overspeed. Fucking shit! I have speed brakes. Everything has gone, all instruments went to shit. Everything has gone. All of them . . .

The captain has deployed the air brakes, but the instruments do not show the airplane slowing down. Rather, they show an increase *in speed, while the aircraft in fact has slowed to near stall speed. The overspeed warning is sounding in his ears while the stick shaker on the control yoke is telling him that he is in a stall.*

CABIN: [*Sound of stall-warning alarm*]
COPILOT: We are going down!
PILOT: Ah . . .
COPILOT: I don't think so. It can't be overspeed.
CABIN: [*Sound of stall-warning alarm*]
COPILOT: We are flying. . . .
CABIN: [*Sound of stall-warning alarm*]
COPILOT: [*To ATC*] Is there any possibilities of—
CABIN: [*Sound of stall-warning alarm*]
COPILOT: [*To ATC*] We still have overspeed.
PILOT: But with spoilers. All of the—
ATC: Correct, rescue has been warned.
COPILOT: [*To ATC*] We request . . . Is there any airplane that can take off to rescue us?
PILOT: Eh? Wait, no, no, no, no.

ATC: Yes, correct. We are going to coordinate immediately. It's being coordinated immediately.

COPILOT: [*To ATC*] Any plane that can guide us, an AeroPeru that may be around? Somebody?

PILOT: Don't tell him anything about that!

COPILOT: Yes, because right now we are stalling.

ATC: Attention, we have a [Boeing] Seven oh seven that will depart. We are telling him.

PILOT: We are not stalling! It's fictitious, it's fictitious.

The speedometers indicate overspeed, which the pilot believes to be true. Meanwhile, the airbrakes are stalling the airplane. The captain has two choices, and he chooses to believe what the instruments are telling him, unaware that the instruments are unable to give an accurate reading. The copilot, on the other hand, believes what the stick shaker tells him—that they are in a stall.

COPILOT: No! If we have shaker how would it be not—

PILOT: Shaker . . . But it is . . . Eh, but even with speed brakes on we are maintaining nine five zero feet. . . . Why do we read the same? I don't understand. . . . Power. What power do we have?

The pilot continues to believe his instruments, now his altimeter. But throttle settings do not coincide with the speed that the instruments are indicating—454 mph.

COPILOT: It's coming back, look. My speedometer is coming back.

PILOT: No, it has gone to off.

COPILOT: It has gone to off.

ATC: AeroPeru Six zero three, you have turned slightly to the left. Now you are heading three two zero and your level is one hundred, approximate speed of two twenty knots and a distance of thirty-two miles northwest of Lima.

CABIN: [*Sound of double-alert alarm*]

COPILOT: You are going down! Eh . . . one zero zero to runway, look! Let's program. I think . . . eh.

COPILOT: [*To ATC*] We have a problem here.

COPILOT: Overspeed. It doesn't cut. The overspeed . . . Where is the overspeed? Where is the overspeed warning? Shit!

PILOT: Nine thousand feet if it indicates you. . . . It's fictitious. Everything is fictitious. The air data has gone to shit.

ATC: AeroPeru Six zero three, your base is [asking] if you have both computer systems out of service.

Copilot: [*To ATC*] We have . . . No instrument is working as speedometer, we have overspeed alarm, and no . . . Apparently it is not. Having overspeed shaker. The fast forward is indicating to high speed. We have engines cut down and [it] doesn't decelerate apparently.

Pilot: Apparently.

Copilot: Let's see. Accelerate one engine. No, it ejects it back.

Pilot: Because the auto-throttle is on.

Copilot: Auto-throttle off. Let's see. Accelerate!

ATC: The Seven oh seven will be ready in some fifteen minutes to fly west in your help.

Cabin: [*Sound of mechanical alert: "Too low terrain" from ground-proximity warning system (G.P.W.S.)*]

Pilot: What happen?

Copilot: Too low terrain.

The crew could refer to radio altimeters, which do not depend on the static ports. However, they choose to rely on the altimeters that require the static ports, thereby worsening their predicament.

Pilot: But . . . Let's go left.

The pilot assumes that they are flying over the ground, rather than over water.

Copilot: [*To ATC*] We have terrain alarm, we have terrain alarm!

Cabin: [*Sound of wind-shear warning: "Too low terrain."*]

ATC: Roger, according to monitor, it indicates flight level one zero zero, over the sea, heading a northwest course of three zero zero.

Copilot: [*To ATC*] We have terrain alarm and we are supposed to be at ten thousand feet?

Pilot: Shit, we have everything.

Cabin: [*Sound of wind-shear warning*]

ATC: According to monitor, you have one zero five.

Copilot: [*To ATC*] We have all computers crazy here.

Pilot: Shit, what the hell these assholes have done?

ATC: Roger, AeroPeru Six zero three, it shows that you are turning left, you are aiming to the west [out to sea].

COPILOT: [*To ATC*] I don't understand.
ATC: It shows that you are turning left, you are aiming to the west.
COPILOT: [*To ATC*] Affirmative, we are heading two five zero, but we are going over the sea because we have low-terrain alarm.
PILOT: We are going to the sea.
ATC: Yes, affirmative. It shows that you are at forty-two miles, flying with course west, course two five zero heading west.
COPILOT: [*To ATC*] We are over water, aren't we?
ATC: Affirmative, over the water, you are forty-two miles west.
COPILOT: [*To ATC*] Okay.
PILOT: Are we going down now?
COPILOT: [*To ATC*] We don't have . . . We have like three seven zero knots. Are we descending now?
ATC: It shows the same speed. You have two zero zero speed approximately.
COPILOT: [*To ATC*] Two hundred of speed?
ATC: Two twenty of speed over the ground, reducing speed slightly.
PILOT: Shit! We will stall now.
CABIN: [*Sound of sink-rate alarm*]

A steep descent activates the sink-rate alarm.

COPILOT: Let's go up, let's see. Let's go up here.

Aircraft climbs to 2,400, descends to 1,300, then climbs again to 4,000 feet. But the altimeters show them at 9,000. The crew believes they are flying at 10,000 feet.

COPILOT: Overspeed. This shit.
PILOT: The overspeed no, this . . . eh . . . Cabin altitude alert, Mach, now we must—
COPILOT: Shit.
PILOT: Eh . . . Read one by one, I'll fly here. . . . And forty-five miles . . . You read. You make the emergency.
COPILOT: Overspeed . . . Fuck . . . Go higher.
PILOT: These sons of a bitch.
COPILOT: Your speaker, let's see. . . . Overspeed again. I don't have overspeed. I don't have overspeed. . . . Auto-throttle, cabin altitude inoperative.
ATC: AeroPeru Six zero three, you are at fifty miles from Lima flying west, course two seven zero, with level one zero.

PILOT: Okay . . . Let's go back, let's set course hundred. Hundred . . .

Now the captain sets a course to return to Lima, and he begins to prepare the aircraft for descent and landing.

COPILOT: [*To ATC*] Ah, affirmative, we are setting course three six zero.
ATC: Okay. Roger, correct.
COPILOT: [*To ATC*] Ah, do you have [our] speed?
ATC: Yes, it shows a speed of approximately three hundred knots.
COPILOT: [*To ATC*] What about some plane that can give us [altitude and speed indications]?
ATC: Correct. A Boeing Seven oh seven has been coordinated. It will be ready in about fifteen minutes, the Tower informed some minutes ago.
COPILOT: Autopilot, overspeed, I don't find overspeed.

Copilot is looking in the Boeing operating manual for overspeed instructions.

PILOT: How do we know the speed we have?
COPILOT: Fast, fast. Horizons are okay. . . . Horizons are right. It's the only thing that is right.
PILOT: Flight level change.
COPILOT: [*Reading from manual*] "APU engines, auto-flight, APU engines, auto-flight instruments, cargo fire, ditching, emergency equipment, passenger evacuation, electrical, flight controls, fuel, hydraulic, ice and rain, landing gear, flight instruments, auto-throttle disconnect" . . . Let's see, if we put the—
PILOT: Cabin alt[itude] inoperative.
COPILOT: Yeah, never mind. Let's put in manual to go down manual. Zero ten. That's it. Auto-disconnect . . . Three ninety knots, says here, eh . . . on the airspeed indicator. I'll try to intercept the ILS. I'm trying to go down. [*To ATC*] Lima, AeroPeru Six zero three, we will try to intercept the ILS. Let's see if you tell us if we are in—
CABIN: [*Sound of caution alarm*]
ATC: Roger, AeroPeru Six zero three, you show now level nine seven hundred.
COPILOT: That is right.

ATC: Stand by to verify speed. The [Boeing] Seven oh seven is about to take off. It is starting to move.
Pilot: This is . . . Fuck off.
Copilot: [*To ATC*] Verify the speed. It is very important. We don't have speed on board.
ATC: Correct. You are starting turn, and it shows a velocity of two seventy ground speed.
Copilot: [*To pilot*] Not so much, there it is. Two seventy is fine.

Copilot is asking the captain not to increase the speed.

Pilot: Alerts, altitude alerts.
Copilot: Altitude, autopilot. Don't put any autopilot, Captain. Alt alert, autopilot disconnect.
Pilot: Let's go to—
Copilot: Autopilot disconnect, zero six zero three.
Pilot: Ask him to tell us—
Copilot: [*Reads from manual*] "Autopilot disconnect . . . crew awareness." Nothing else says . . . "Autopilot inoperative, autothrottle disconnect . . . crew awareness. Alt alert, cabin inoperative, rudder ratio," there is nothing new. But the overspeed . . .
ATC: Course to intercept localizer. Approximately at thirty miles. You must fly heading zero seven zero approximately. Keep the course stable.
Copilot: Yes, turn now . . . zero seventy.
Copilot: [*To ATC*] We will keep zero seven zero. We keep course. It seems that we have the right course. What we don't have is airspeed indicator and altitude nine seven hundred.
Pilot: Flight-level change.
ATC: Correct. Altitude is nine seven hundred. And the speed is two forty knots over the ground, according to monitor.
Cabin: [*Sound of warning alarm*]
Copilot: What if we put [down] flaps? . . . Two hundred forty.
Pilot: What?
Copilot: If we put [down] flaps?
ATC: . . . And at fifty-one miles from Lima.
Pilot: Okay.
Copilot: Zero seventy . . . I want to take that shit out of there!
Pilot: The overspeed? But you can't.
Cabin: [*Alarm warning*]

COPILOT: You can't?
PILOT: Let's see. . . . You switch your thing . . .
COPILOT: Air data? No . . . But it's flying well, I have the . . . If not, I'm going . . . Shit.
PILOT: How can it be flying at this speed if we are going down with all the power cut off?
COPILOT: [*To ATC*] Can you tell me the altitude, please, because we have the climb that doesn't—
PILOT: Nine.
CABIN: [*Alarm warning: "Too low terrain."*]
ATC: Yes, you keep nine seven hundred [9,700 feet] according to presentation [on radar], sir.
COPILOT: [*To ATC*] Nine seven hundred?
ATC: Yes, correct. What is the indicated altitude on board? Have you any visual reference?
COPILOT: [*To ATC*] Nine seven hundred, but it indicates too low terrain. . . . Are you sure that you have us on the radar at fifty miles?
PILOT: Hey, look. . . . With three seven zero we have—
COPILOT: Have what? Three seven zero of what? Do we lower the gear?
PILOT: But what do we do with the gear? Don't know . . . that.
CABIN: [*Sound of initial impact with the water*]
COPILOT: [*To ATC*] We are impacting water! Pull it up!!
ATC: Go up, go up if it indicates pull up.
PILOT: I have it, I have it!
CABIN: [*Warning sound: "Too low terrain."*]
PILOT: We are going to invert [turn upside down]!
CABIN: [*Sound of alarms: "Whoop . . . Whoop . . . Pull. . . ."*]
CABIN: [*Sound of impact*]

END OF TAPE

At impact with the water, the onboard altimeters indicated an altitude of 9,700 feet. The aircraft's left wing and engine made initial contact with the water at a ten-degree angle, at 300 mph. The aircraft then climbed 200 feet, turned over, and fell again, crashing into the sea. The wreckage was subsequently recovered. Nine crew members and sixty-one passengers were lost.

3

Ban Nong Waeng, Thailand

May 26, 1991

■

Lauda Air Flight 004

This CVR transcript documents a freak accident. No piloting skills could have saved the 213 passengers and the crew of 10 that night aboard the Lauda Air Flight 004, a Boeing 767, named the *Mozart*, out of Bangkok International Airport, bound for Vienna's Wien-Schwechat. The passengers were mostly Austrian, German, Swiss, and Hong Kong Chinese. Many of them were returning from holidays in Thailand; some were traveling on business.

The flight, which had originated in Hong Kong, took off from Bangkok at 11:10 P.M. Twelve minutes later, the crew got a visual advisory warning on its annunciator panel that a system failure might cause the in-flight deployment of the number-one engine's thrust reverser, which brakes the aircraft on the ground by throwing the thrust of the jet engines forward of the wing through the mechanism of mechanical scoops. A deployment in flight of the thrust reverser is by definition catastrophic. It is almost unthinkable.

The crew assumed that the annunciator warning was a false reading caused by moisture in the system. The crew could not have seriously believed that the thrust reversers would activate in flight, no matter what. Besides, the Boeing 767's Emergency/Malfunction Checklist, to which the crew referred right away, reported, ''No action required.''

Nine minutes after the advisory, twenty-one minutes into the flight, the number-one thrust reverser deployed. The Boeing disintegrated at 4,000 feet.

We pick up the CVR just as the captain receives the notification of "an additional system failure [which] might cause in-flight deployment of the thrust reverser." This was the first and only warning.

CAPTAIN: Shit. That keeps . . . that's coming on.
COPILOT: So we passed transition altitude one zero one three.
CAPTAIN: Okay. What's it say in there about that [deployment in flight of thrust reverser], just ah . . .
COPILOT: [*Reads from Boeing 767 manual*] "Additional system failures may cause in-flight deployment. Expect normal reverse operation after landing."
CAPTAIN: Okay. Just, ah, let's see. Okay.
COPILOT: Shall I ask the ground staff?
CAPTAIN: What's that?
COPILOT: Shall I ask the technical men [on the ground]?
CAPTAIN: Ah, you can tell 'em it, just it's, it's, it's, just—ah, no, ah, it's probably, ah, well . . . ah, moisture or something 'cause it's not just, oh, it's [the annunciator panel light is] coming on and off.
COPILOT: Yeah.
CAPTAIN: But, ah, you know it's a . . . it doesn't really, it's just an advisory thing, I don't, ah . . . Could be some moisture in there or somethin'.
COPILOT: [I] think you need a little bit of rudder trim to the left.
CAPTAIN: What's that?
COPILOT: You need a little bit of rudder trim to the left. [*He starts adding up figures in German*]
CAPTAIN: [*Apparently sees instruments*] Ah, reverser's deployed.
CABIN: [*Sound of snap*]
CAPTAIN: Jesus Christ!
CABIN: [*Sound of four caution tones; sound of siren warning starts; sound of siren warning stops; sound of siren warning starts and continues until the recording ends*]
CAPTAIN: Here, wait a minute! Damn it!
CABIN: [*Sound of bang*]

END OF TAPE

Lauda Air Flight 004 disintegrated at an elevation of nearly 4,000 feet; the fuselage crashed in the Thai jungle at a 45-degree angle.

The cockpit and one of the aircraft's wings were found more than a mile from the main crash site. The probable cause of the crash was the in-flight deployment of the number-one engine thrust reverser, which itself was caused by a failure of the thrust reverser isolation valve. The official report determined that "recovery from the event was uncontrollable [*sic*] for an unexpecting flight crew." All aboard were lost.

San Diego, California

September 25, 1978

■

Pacific Southwest Airlines Flight 182

That morning in clear California skies, Pacific Southwest Airlines Flight 182, a Boeing 727-214 with 7 crew members and 128 passengers, was making its approach into San Diego's Lindbergh Field after a short flight that had originated at Los Angeles International Airport. The approach was visual to Runway 27, over the office towers on the hills of San Diego that overlook the harbor.

The PSA crew was aware of other airplane traffic in the area, including the presence of a small, single-engined Cessna 172, which had taken off at 8:16 A.M. from nearby Montgomery Field and had proceeded to Lindbergh Field, where it had already practiced two instrument landings. At 09:00 A.M. the Cessna pilot was instructed to maintain Visual Flight Rules at or below 3,500 feet, heading 70 degrees. At that moment, PSA Flight 182 was flying over San Diego at 2,600 feet.

We pick up the CVR tape just as the PSA cockpit is being notified of traffic in their area.

APPROACH CONTROL: PSA One eighty-two, traffic [at] twelve o'clock, one mile northbound.

CAPTAIN: We're looking.

APPROACH: PSA One eighty-two, additional traffic's, ah, twelve o'clock, three miles just north of the field northwestbound, a Cessna One seventy-two climbing VFR [Visual Flight Rules] out of one thousand four hundred [feet].

COPILOT: Okay, we've got that other [at] twelve [o'clock].

APPROACH: [*To the Cessna 172*] Cessna Seven seven one one Golf, San Diego Departure radar contact, maintain VFR conditions at or below three thousand five hundred, fly heading zero seven zero, vector final approach course.
APPROACH: PSA One eighty-two, traffic's at twelve o'clock, three miles out of one thousand seven hundred.
COPILOT: Got 'em.
CAPTAIN: Traffic in sight.
APPROACH: Okay, sir, maintain visual separation, contact Lindbergh Tower one three three point three, have a nice day now.
CAPTAIN: Okay.
CAPTAIN: Lindbergh, PSA One eighty-two downwind.
LINDBERGH TOWER: PSA One eighty-two, Lindbergh Tower, ah, traffic twelve o'clock one mile a Cessna.
COPILOT: [*To captain*] Flaps five.
CAPTAIN: Is that the one [the traffic] we're looking at?
COPILOT: Yeah, but I don't see him now.
CAPTAIN: Okay, we had it there a minute ago.
TOWER: One eighty-two, roger.
CAPTAIN: I think he's [the traffic has] pass[ed] to our right.
TOWER: Yeah.
CAPTAIN: He was right over here a minute ago.
TOWER: How far are you going to take your downwind? Company traffic is waiting for departure.
CAPTAIN: Ah, probably about three to four miles.
TOWER: Okay. PSA One eighty-two, cleared to land.
CAPTAIN: One eighty-two's cleared to land.
COPILOT: Are we clear of that Cessna?
FLIGHT ENGINEER: Suppose to be.
CAPTAIN: I guess.
A PSA CAPTAIN: [*Who was standing in cockpit, observing*] I hope.
CAPTAIN: Oh, yeah, before we turned downwind, I saw him about one o'clock. [He's] probably behind us now.
COPILOT: There's one [an airplane] underneath [us]. I was looking at that inbound there.
CAPTAIN: Whoop!
COPILOT: Aghhh!
CABIN: [*Sound of impact with the Cessna in midair*]
CAPTAIN: Easy, baby, easy, baby. What have we got here?
COPILOT: It's bad. We're hit, man, we are hit.

CAPTAIN: Tower, we're going down, this is PSA. . . .
TOWER: Okay, we'll call the equipment for you.
CABIN: [*Sound of stall warning*]

END OF TAPE

After collision with the Cessna, the PSA plummeted to the ground into a residential area of San Diego. All 135 passengers and crew of the PSA died; 9 people on the ground were also killed.

Sakhalin, Russia

September 1, 1983

■

Korean Air Flight 007

The story of Korean Air Flight 007 (KAL 007) is by now part of the dark legend of the Cold War. Mistaken as a U.S. spy plane, the Korean Air Boeing 747, which was flying well off course into restricted Soviet airspace, was approached and shot down by an intercepting MIG pilot, who had sighted KAL and fired a heat-seeking missile at its fuselage. Out of control, KAL plunged to the ocean below, where its black box was recovered weeks later.

We pick up the transcript seconds before the missile struck the airplane and follow for roughly seven minutes as the Boeing 747 plunged 35,000 feet into the sea.

TOKYO AIR-TRAFFIC CONTROL: Korean Air Zero zero seven clearance, Tokyo ATC clears Korean Air Zero zero seven [to] climb and maintain flight level three five zero [35,000 feet].

CAPTAIN: Ah, roger, Korean Air Zero zero seven, climb and maintain at three five zero, leaving three three zero at this time [33,000 feet].

CONTROL: Tokyo roger.

CABIN: [*Sound of altitude alert*]

CAPTAIN: Tokyo radio, Korean Air Zero zero seven reaching level three five zero [35,000 feet].

CONTROL: Korean Air Zero zero seven, Tokyo roger.

CABIN: [*Sound of explosion*]

CAPTAIN: What's happened?

COPILOT: What?

CAPTAIN: Retard throttles.
COPILOT: Engines normal.
CAPTAIN: Landing gear.
CABIN: [*Sound of cabin-altitude warning*]
CAPTAIN: Landing gear.
CABIN: [*Sound of altitude-deviation warning; sound of autopilot-disconnect warning*]
CAPTAIN: Altitude is going up. Altitude is going up. Speed brake is coming out.
COPILOT: What? What?
CAPTAIN: Check it out.
CABIN: [*Sound of public-address-system chime for automatic cabin announcement*]
COPILOT: I am not able to drop altitude; now unable.
CABIN: [*Public-address recording: "Attention, emergency descent."*]
CAPTAIN: Altitude is going up. This is not working. This is not working. Manually.
COPILOT: Cannot do manually.
CABIN: [*Public-address recording (in Japanese): "Attention, emergency descent." Sound of autopilot-disconnect warning*]
COPILOT: Not working manually also. Engines are normal, sir.
CABIN: [*Public-address recording: "Put out your cigarette. This is an emergency descent. Put out your cigarette. This is an emergency descent."*]
CAPTAIN: Is it power compression?
FLIGHT ENGINEER: Is that right?
CABIN: [*Public-address recording: "Put out your cigarette. This is an emergency descent."*]
FLIGHT ENGINEER: All of both . . .
CAPTAIN: Is that right?
CABIN: [*Public-address recording: "Put the mask over your nose and mouth and adjust the headband."*]
COPILOT: Tokyo radio, Korean Air Zero zero seven.
CABIN: [*Public-address recording: "Put the mask over your nose and mouth and adjust the headband."*]
CONTROL: Korean Air Zero zero seven, Tokyo.
COPILOT: Roger, Korean Air Zero zero seven . . . ah, we are experiencing . . .
CABIN: [*Public-address recording: "Put the mask over your nose and mouth and adjust the headband."*]
FLIGHT ENGINEER: All compression.

Captain: Rapid decompression. Descend to one zero thousand [10,000 feet].

Cabin: [*Public-address recording: "Attention, emergency descent. Attention, emergency descent."*]

Flight Engineer: Now . . . we have to set this.

Tokyo: Korean Air Zero zero seven, radio check on one zero zero four eight.

Cabin: [*Public-address recording: "Attention, emergency descent."*]

Flight Engineer: Speed. Stand by. Stand by. Stand by. Stand by. Set.

Cabin: [*Public-address recording: "Put out your cigarette. This is an emergency descent. Put out your cigarette. This is an emergency descent. Put out your cigarette. This is an emergency descent. Put the mask over your nose and mouth and adjust the headband. Put the mask over your nose and mouth and adjust—"*]

END OF TAPE

All aboard were lost.

Indian Ocean Descent

November 28, 1987

■

South African Airways Flight 295

As much as anything else, South African Airways Boeing 747-244B, registered as Flight 295, which was carrying 140 passengers and a crew of 19, shows the confusion that can overcome a cockpit crew when confronted with an emergency—in this instance, a catastrophic cargo-hold fire.

Flight 295 took off from Chiang Kai-shek Airport in Taiwan that afternoon, bound for South Africa. Just before midnight, while flying over the southern Indian Ocean, the crew reported a fire in one of the six cargo pallets in the main air-freight hold. The Boeing 747 made an emergency descent, and Mauritius Air-Traffic Control gave the flight an approach clearance.

We pick up the CVR transcript just as the cockpit crew notifies Mauritius Approach of a fire on board and asks for permission to descend to 14,000 feet.

CAPTAIN: [*To Mauritius Approach*] Er, good morning. We have, er, a smoke problem and we are doing an emergency descent to level one five, er, one four zero [14,000 feet].

MAURITIUS APPROACH: Confirm you wish to descend to flight level one four zero?

CAPTAIN: *Ja*, we have already commenced, er, due to a smoke problem in the airplane.

APPROACH: Eh, roger. You are clear to descend immediately to flight level one four zero.

CAPTAIN: Roger, we will appreciate it if you can alert, er, fire, er, er, er.

APPROACH: Do you request a full emergency, please? A full emergency?

CAPTAIN: Affirmative, that's Charlie, Charlie.

APPROACH: Roger, I declare a full emergency.

CAPTAIN: Thank you.

APPROACH: [*Asks the aircraft crew for a position report*]

CAPTAIN: Now we have lost a lot of electrics. We haven't got any [lights] on the aircraft now.

APPROACH: [*Asks for an estimated time of arrival and an update of the aircraft's position; advises that both runways are available*]

CAPTAIN: Er, we'd like to track in, er, One, er, One three [Runway 13].

APPROACH: Confirm Runway Thirteen?

CAPTAIN: Charlie, Charlie [yes, yes].

CABIN: [*Fire alarm bell sounds, followed by public-address-system chime*]

FLIGHT ENGINEER: What's going on now . . . ? Cargo? It came on now afterwards. Main-deck cargo . . . then the other one came on as well. I've got two.

CAPTAIN: [*Calls for checklist to be read*] It is the fact that both came on. It disturbs one.

VOICE IN COCKPIT: Aagh!

CAPTAIN: What's going on now?

CABIN: [*Sudden bang*]

END OF TAPE

The probable cause of the crash was a fire of unknown origin that had possibly: 1) incapacitated the crew; 2) caused disorientation of the crew due to thick smoke; 3) caused crew distraction; 4) weakened the aircraft structure, causing an in-flight breakup; 5) burned through several control cables; 6) caused loss of control due to deformation of the aircraft fuselage.

All on board were lost.

Cove Neck, New York

January 25, 1991

■

Avianca Flight 052

This disaster has been much analyzed for what the cockpit crew did and did not do in a timely fashion.

The captain of Avianca Flight 052, a Boeing 707 originating in Medellín, Colombia, bound for New York's JFK Airport, had repeatedly asked his copilot to declare a "fuel emergency" with the air-traffic controllers at JFK. The fuel aboard the 707 was unexplainably low upon arrival in the New York City area. The flight crew, long aware of the fuel situation, had tried to maximize the efficiency of the fuel pumps, so that the last few drops of jet fuel in the tanks would reach the four Pratt & Whitney engines. A declaration of a fuel emergency to New York Approach or the Kennedy Tower would have given Flight 052 priority over all other aircraft and without a doubt would have taken it out of harm's way. Without an emergency declaration, and with the Approach controllers unaware of the plane's urgent need to land, Flight 052 circled JFK for nearly an hour. The captain was flying the aircraft, the copilot communicating with New York Approach and the Tower. The time was around 9:00 P.M.

COPILOT: New York Approach, Avianca Zero five two leveling five thousand [feet].

NEW YORK APPROACH: Avianca Zero five two heavy, New York Approach, good evening. Fly heading zero six zero.

COPILOT: Heading zero six zero, Avianca Zero five two heavy.

FLIGHT ENGINEER: When we have . . . a thousand pounds or less in any tank, it is necessary to do [an emergency landing].

COPILOT: Yes, sir.

FLIGHT ENGINEER: The go-around procedure is stating that the power be applied slowly and to avoid rapid accelerations and to have a minimum of nose-up attitude.

CAPTAIN: To maintain what?

COPILOT: Minimum, minimum nose-up attitude. That means the less nose-up attitude that one can hold.

FLIGHT ENGINEER: This thing is going okay.

COPILOT: Then flaps to twenty-five position and maintain Vee ref plus twenty. . . . The highest go-around procedure is starting.

FLIGHT ENGINEER: [*Reading from flight manual*] "The flaps"—sorry—"retract the landing gear with positive rate of climb. . . . If any low-pressure light comes on do not select the switch in the off position. . . ." The low-pressure lights of the [fuel] pumps comes on. Reduce the nose-up attitude, the nose-up attitude.

COPILOT: The forward [fuel] pumps . . .

CAPTAIN: What heading do you have over there? Select Kennedy on my side.

COPILOT: Kennedy is on the number two, but if [you] want, Commander, I can perform the radio setup right now that we are now being vectored. We are like on downwind position now.

CAPTAIN: We passed [the airport] already, no?

COPILOT: Yes, sir.

APPROACH: Avianca Zero five two heavy, turn left, heading three six zero.

COPILOT: Left, heading three six zero, Avianca Zero five two heavy.

ENGINEER: Three six zero.

COPILOT: Yes, Commander, that's what he says.

CAPTAIN: Perform the radio setup, but leave to me the VOR. . . . Two what?

COPILOT: Two twenty-three.

CAPTAIN: Two twenty-three. What heading he provide us?

COPILOT: New. He give us three six zero.

CAPTAIN: Okay.

COPILOT: I am going to perform the radio setup on number two.

CAPTAIN: Perform the radio setup.

CABIN: [*Sound of landing-gear warning horn*]

CAPTAIN: Hey, [I] understand that the nose must be maintained as low as possible, yes?

FLIGHT ENGINEER: That's correct. It says that the forward pumps . . .

The flight engineer is trying to keep fuel supplied to the engines. The "pumps" he is talking about are pumping fuel into the engines.

APPROACH: Avianca Zero five two heavy, turn left heading of three zero zero.

COPILOT: [*To Approach*] Left heading three zero zero, Avianca Zero five two heavy. Three zero zero on the heading.

FLIGHT ENGINEER: The forward boost pumps could be uncovered on fuel during the go-around. What it means it doesn't contain fuel for feeding itself and a flameout can occur [in other words, the pumps may have no fuel to pump] and it is necessary to lower the nose again [to get fuel into the pumps].

CAPTAIN: Heading three hundred.

COPILOT: Three hundred. Right now we are proceeding to the airport inbound and we have twenty-seven, seventeen miles.

FLIGHT ENGINEER: Roger.

COPILOT: This means that we'll have hamburger tonight.

APPROACH: Avianca Zero five two heavy, turn left heading two niner.

COPILOT: Left heading two niner zero, Avianca Zero five two heavy. Two niner zero on the heading, please.

CAPTAIN: Two twenty-three course counter, stand by the frequency number.

COPILOT: Stand by for the frequency.

CAPTAIN: Leave the ILS frequency in Kennedy until I advise you. Select your own there.

COPILOT: It is ready.

CAPTAIN: Well . . .

COPILOT: Markers are set.

APPROACH: Avianca Zero five two heavy, descend and maintain, ah, descend and maintain three thousand [feet].

COPILOT: Descend and maintain three thousand, Avianca Zero five two heavy.

FLIGHT ENGINEER: They got us. They [are] already vectoring us.

COPILOT: They accommodate us [are letting us land] ahead of an—
CAPTAIN: What?
COPILOT: They accommodate us.
FLIGHT ENGINEER: They already know that we are in bad condition.
CAPTAIN: No, they are descending us.
COPILOT: One thousand feet.
CAPTAIN: Ah, yes.
COPILOT: They are giving us priority.
APPROACH: Avianca Zero five two heavy, turn left heading two seven zero.
COPILOT: Left heading two seven zero.
COPILOT: Two seven zero on the heading.
CAPTAIN: Two seventy.
COPILOT: [The runway] is ahead of us.
FLIGHT ENGINEER: Yes.
CAPTAIN: Stand by for the localizer there.
COPILOT: Yes, sir. Outer marker is seven miles.
APPROACH: Avianca Zero five two heavy, turn left heading two five zero, intercept the localizer.
COPILOT: Heading two five zero, intercept the localizer, Avianca Zero five two heavy. This is final vector. Do you want the ILS, Commander?
APPROACH: Avianca Zero five two heavy, you are one five miles from the outer marker. Maintain two thousand [feet] until established on the localizer. Cleared ILS [Runway] Twenty-two left.
FLIGHT ENGINEER: Two thousand.
CAPTAIN: Select the ILS on my side.
COPILOT: The ILS in number one, one hundred ten point nine, is set for two thousand feet.
FLIGHT ENGINEER: Localizer alive.
CAPTAIN: Give me flaps fourteen.
COPILOT: We are thirteen miles from the outer marker. Flaps fourteen.
CAPTAIN: Navigation number one. Did you already select flaps fourteen, no?
COPILOT: Yes, sir, [they] are set. Navigation number one.
APPROACH: Avianca Zero five two heavy, speed one six zero, if practical.
CAPTAIN: Give me flaps twenty-five.

COPILOT: We have traffic ahead of us.
CAPTAIN: We can maintain one hundred and forty [mph] with this flap setting. How many miles is that thing located?
COPILOT: It is at seven miles, Commander, and we are at ten miles at the moment from the outer marker.
CAPTAIN: Reset frequency, the ILS, please.
COPILOT: Now the course is going to be intercepted at the outer marker. This means there is not a problem, Commander. Localizer to the left.
APPROACH: Avianca Zero five two heavy, contact Kennedy Tower, one one niner point one, good day.
COPILOT: One one niner point one, so long. Kennedy Tower, Avianca Zero five two established Twenty-two left.
KENNEDY TOWER: Avianca Zero five two heavy, Kennedy Tower, Twenty-two left. You're number three [to land] following [a Boeing] Seven two seven traffic on, ah, niner mile final.
COPILOT: Avianca Zero five two, roger.
CAPTAIN: Can I lower the landing gear yet?
COPILOT: No, I think it's too early now. If we lower the landing gear, we have to hold very high nose attitude.
FLIGHT ENGINEER: And it's not very . . .
TOWER: Avianca Zero five two, what's your airspeed?
COPILOT: Avianca Zero five two, one four zero knots.
CAPTAIN: They were asking for the American [Airlines 727 that was preceding the Avianca in for a landing].
TOWER: Avianca Zero five two, can you increase your airspeed one zero knots?
CAPTAIN: One zero.
COPILOT: Okay, one zero knots, increasing.
TOWER: Increase, increase!
CAPTAIN: What?
COPILOT: Increasing.
CAPTAIN: What?
TOWER: Okay.
FLIGHT ENGINEER: Ten knots more.
COPILOT: Ten little knots more.
CAPTAIN: Here we go. Tell me things louder, because I'm not hearing it. [The captain is hard of hearing.]
COPILOT: We are three miles to the outer marker now.
CAPTAIN: Right resetting the ILS.

COPILOT: Here it is already intercepted. Glide slope alive.
CAPTAIN: I'm going to approach at one hundred and forty [mph]. It is what he wants, or what is the value he wants?
COPILOT: One hundred and fifty. We had one hundred and forty, and he required ten little knots more.
CAPTAIN: Lower the gear.
COPILOT: Gear down.
CAPTAIN: Mode selector approach land checklist.

They go through the landing checklist.

TOWER: Avianca Zero five two, Two two left, wind one niner zero at two zero. Cleared to land.
FLIGHT ENGINEER: [Landing] Gear.
COPILOT: Cleared to land, Avianca Zero five two heavy. Wind check, please?
TOWER: One niner zero at two zero.
COPILOT: Thank you. One hundred and ninety with twenty is in the wind.
CAPTAIN: With what?
TOWER: Avianca Zero five two, say airspeed?
COPILOT: Zero five two is, ah, one four five knots.
TOWER: TWA Eight oh one heavy, if feasible reduce airspeed one four five.
CAPTAIN: Give me fifty [flaps]. Are we cleared to land?
COPILOT: Yes, sir, we are cleared to land.
FLIGHT ENGINEER: Hydraulic pressure quantities normal.
COPILOT: Stand by flaps fifty.
CAPTAIN: Give me fifty.
FLIGHT ENGINEER: All set for landing.
COPILOT: Below glide slope.
CAPTAIN: Confirm the wind.
TOWER: Avianca Zero five two heavy, can you increase your airspeed one zero knots at all?
COPILOT: Yes, we're doing it.
TOWER: Okay, thank you.
CAPTAIN: Confirm the wind.

For the next minute and a half, the Avianca flight follows its glide path toward Kennedy International.

CABIN: [*Ground-proximity warning system sounds: "Whoop, whoop! Pull up! Pull up!"*]
COPILOT: Sink rate. Five hundred feet.
CAPTAIN: Lights. Where is the runway? The runway! Where is it?
CABIN: [*Ground-proximity warning system sounds: "Glide slope."*]
COPILOT: I don't see it! I don't see it!
CAPTAIN: Give me the landing gear up. Landing gear up.
CABIN: [*Sound of landing-gear warning horn*]
CAPTAIN: Request another traffic pattern.
COPILOT: [*To Tower*] [We are] executing a missed approach, Avianca Zero five two heavy.
FLIGHT ENGINEER: Smooth with the nose, smooth with the nose, smooth with the nose.
TOWER: Avianca Zero five two heavy, roger. Ah, climb and maintain two thousand, turn left, heading one eight zero.

At this point, the Avianca flight, having missed its approach, goes around.

CAPTAIN: [*To cockpit crew*] We don't have the fuel [to go around]. . . .
COPILOT: Maintain two thousand feet, one eight zero on the heading.
CAPTAIN: Flaps twenty-five. I don't know what happened with the runway. I didn't see it.
FLIGHT ENGINEER: I didn't see it.
COPILOT: I didn't see it.
TOWER: Avianca Zero five two, you are making a left turn, correct, sir?
CAPTAIN: Tell them we are in emergency.
FLIGHT ENGINEER: Two thousand feet.
COPILOT: [*To Tower*] That's [turn] right to one eight zero on the heading, and, ah, we'll try [to land] once again. We're running out of fuel.

The copilot mentions to the Tower that they are low on fuel, but he does not *declare an emergency; the declaration itself must be made in order for the declaring flight to be given emergency priorities to land.*

TOWER: Okay.
CAPTAIN: What did [Kennedy Tower] say?

COPILOT: Maintain two thousand feet, one eight on the heading. I already advise him that we are going to attempt again, because we now can't.

CAPTAIN: Advise [Kennedy Tower] we are [in an] emergency! Did you tell him?

COPILOT: Yes, sir. I already advised him.

CAPTAIN: Flaps four . . . fifteen.

TOWER: Avianca Zero five two heavy, continue the left turn, heading one five zero, maintain two thousand.

CAPTAIN: They [told] us to reduce airspeed, that's the thing, man. Hundred and fifty [mph].

COPILOT: One hundred and fifty on the heading.

CAPTAIN: Flaps fifteen.

COPILOT: [*To Approach*] Approach, Avianca Zero five, ah, two heavy—we just missed missed [our] approach, and, ah, we're maintaining two thousand and five on the . . .

APPROACH: Avianca Zero five two heavy, New York, good evening. Climb and maintain three thousand.

CAPTAIN: Advise him we don't have fuel.

The copilot still does not declare a fuel emergency.

COPILOT: [*To Approach*] Climb and maintain three thousand, and, ah, we're running out of fuel, sir.

APPROACH: Okay, fly heading zero eight zero.

CAPTAIN: [*To copilot*] Did you already advise that we don't have fuel?

COPILOT: Yes, sir, I already advise him. . . . We are going to maintain three thousand feet, and he's going to get us back.

CAPTAIN: Okay.

APPROACH: Avianca Zero five two heavy, turn left, heading zero seven zero.

COPILOT: Heading zero seven zero, Avianca Zero five two heavy.

APPROACH: And, Avianca Zero five two heavy, ah, I'm going to bring you about fifteen miles northeast and then turn you back onto the approach. Is that fine with you and your fuel?

COPILOT: I guess so, thank you very much.

CAPTAIN: What did he say?

FLIGHT ENGINEER: The guy [the controller] is angry.

COPILOT: Fifteen miles in order to get back to the localizer.

CAPTAIN: I'm going to follow this.
COPILOT: We must follow the identified ILS.
CAPTAIN: To die. Take it easy, take it easy.
COPILOT: [*To Tower*] Ah, can you give us a final now? Avianca Zero five two heavy.
APPROACH: Avianca Zero five two, affirmative, sir. Turn left, heading zero four zero.
FLIGHT ENGINEER: I have the [landing] lights on.
APPROACH: [*To TWA Flight 801*] TWA Eight zero one heavy, turn left, heading two five zero, you're one five miles from the outer marker. Maintain two thousand until established on the localizer. Cleared for ILS Two two left.
COPILOT: Avianca Zero five two heavy, left turn two five zero, and, ah, we're cleared for ILS.
CAPTAIN: What heading? Tell me.
COPILOT: Two five zero.
APPROACH: Avianca Fifty-two, climb and maintain three thousand [feet].
CABIN: [*Sound of landing-gear warning horn*]
COPILOT: [*To pilot*] Ah, negative, sir. We [are] just running out of fuel. We're okay three thousand. Now okay.
CAPTAIN: No, no, three . . . three thousand, three thousand.
APPROACH: Okay, turn left, heading three one zero, sir.
COPILOT: Three one zero, Avianca Zero five two.
CAPTAIN: Tell me . . .
COPILOT: Three one zero in the . . .
CAPTAIN: Flaps fourteen.
COPILOT: Three one zero.
FLIGHT ENGINEER: No, sir, are in . . .
CAPTAIN: Set flaps fourteen.
APPROACH: Avianca Fifty-two, fly heading of three six zero, please.
COPILOT: Fourteen degrees.
CAPTAIN: Tell me heading. What?
COPILOT: Okay, we'll maintain three six zero now.
APPROACH: Okay, and you're number two [in line] for the approach. I just have to give you enough room so you can make it without, ah, having to [go around] again.
COPILOT: Okay, we're number two and flying three six zero now.
CAPTAIN: Three sixty, no?

COPILOT: Three sixty.
CAPTAIN: Flaps fourteen.
APPROACH: [*To TWA 801*] TWA Eight zero one heavy, you're eight miles behind a heavy jet [Avianca]. Contact Kennedy Tower, one one niner point one. Thanks for the help.
TWA 801: [*To Approach*] Okay, Eight oh one, roger, and what's his ground . . . what's his airspeed, do you know?
APPROACH: Ah, he's indicating ten knots slower, eight miles.
TWA 801: Okay, thank you.
APPROACH: [*To Pan Am Flight 1812*] Clipper Eighteen twelve heavy, speed one six zero, if practical.
PAN AM 1812: Eighteen twelve heavy slowing to one fifty.
APPROACH: Avianca Zero five two heavy, turn left, heading three three zero.
COPILOT: Three three zero on the heading, Avianca Zero five two. Three three zero, the heading.
FLIGHT ENGINEER: Flameout! Flameout on engine number four.

The tanks are out of fuel.

CABIN: [*Sound of momentary power interruption to the CVR*]
CAPTAIN: Flameout on it [the engine].
FLIGHT ENGINEER: Flameout on engine number three, essential on number two, one number one.
CAPTAIN: Show me the runway.
COPILOT: [*To Approach*] Avianca Zero five two, we just, ah, lost two engines and, ah, we need priority, please.
APPROACH: Avianca Zero five two, turn left, heading two five zero, intercept the localizer.
CABIN: [*Sound of engine spooling down*]
COPILOT: Two five zero. Roger.
CAPTAIN: Select the ILS.
APPROACH: Avianca Zero five two heavy, you're one five miles [fifteen miles] from the outer marker, maintain two thousand [feet] until established on the localizer. Cleared for ILS Two two left.
COPILOT: Roger, Avianca.
CAPTAIN: Did you select the ILS?
COPILOT: It is ready on two.

END OF TAPE

With its engines starved of fuel, the Avianca flight crashed into Cove Neck. The captain, copilot, and flight engineer died, as did five of the six flight attendants. Of the 158 passengers on board, 73 perished, 82 were seriously injured, 3 suffered minor injuries. The aircraft hit the ground at 9:35 P.M.

Puerto Plata, Dominican Republic

February 6, 1996

■

Birgen Air ALW301

For seven minutes after takeoff from Gregorio Luperon Airport, the Birgen Air Boeing 757, charter Flight ALW301 with 176 passengers and a crew of 13, flew into the late-night darkness over the eastern Caribbean, en route to Frankfurt, Germany. During the takeoff roll-out, the pilot's airspeed indicator was inaccurate, and the copilot read out the speeds from the speedometer on his side of the cockpit. The crew saw no reason to abort the flight; however, the airspeed error, which was caused by a blocked pitot port outside the aircraft (similar to the static ports on the doomed AeroPeru flight; see Chapter 2) confused the pilot, who allowed the aircraft to stall. Without recovering, it fell into the sea five miles off the coast of the Dominican Republic and sank in 7,200 feet of water. The flight recorders were recovered in the deep on February 28, 1996, at a cost of $1.4 million.

We pick up the CVR transcript just as the Boeing 757 is accelerating down the airstrip for takeoff.

GREGORIO LUPERON TOWER: Have a nice flight.
CABIN: [*Sound of increasing engine noise*]
COPILOT: Power's set.
CAPTAIN: Okay, checked.
COPILOT: Eighty knots.
CAPTAIN: Checked. My airspeed indicator's not working.
COPILOT: Yes. Yours is not working. One twenty [mph].
CAPTAIN: Is yours working?

COPILOT: Yes, sir.
CAPTAIN: You tell me [the ground speed].
COPILOT: Vee one. Rotate.
CAPTAIN: Positive climb. Gear up.
COPILOT: Positive climb.
CABIN: [*Sound of landing-gear handle being moved*]
COPILOT: Gear is up. LNAV [Lateral Navigation on autopilot]?
CAPTAIN: Yes, please.
COPILOT: LNAV.
CAPTAIN: Yes.
COPILOT: It began to operate.
CAPTAIN: Is it possible to turn off the [windshield] wipers?
COPILOT: Okay, wipers off.
CABIN: [*Sound of windshield wipers stops*]
CAPTAIN: Climb thrust.
COPILOT: Climb thrust.
CAPTAIN: VNAV [Vertical Navigation on autopilot].
COPILOT: VNAV. Okay flap speed.
CAPTAIN: Flaps five, flaps one.
COPILOT: Flaps to one.
CAPTAIN: Gear handle off.
COPILOT: Gear handle's off.
CAPTAIN: Flaps up.
COPILOT: Flaps up.
CAPTAIN: [Begin] after-takeoff checklist.
COPILOT: After-takeoff checklist: landing gear up and off, flaps are up, checked up, altimeters later, after-takeoff completed.
CAPTAIN: Okay. Center autopilot on, please.
COPILOT: Center autopilot is on command.
CAPTAIN: Thank you. One zero one three. Rudder ratio, Mach airspeed trim.
COPILOT: Yes, trim.
CAPTAIN: There is something wrong. There are some problems. Okay, there is something crazy. Do you see it?
COPILOT: There is something crazy there at this moment. Two hundred [mph] only on mine [my speedometer] and [the speed is] decreasing, sir.
CAPTAIN: Both of [the speedometers] are wrong. What can we do? Let's check their circuit breakers.
COPILOT: Yes.

CAPTAIN: Alternate is correct.
COPILOT: The alternate one is correct.
CAPTAIN: As aircraft was not flying and [when it was] on ground, something happening. . . . We don't believe [the speedometers].

Here we listen to the first signs of the pilot's utter confusion.

FLIGHT ENGINEER: Shall I reset its circuit breaker?
CAPTAIN: Yes, reset it.
FLIGHT ENGINEER: To understand the reason?
CAPTAIN: Yeah.
CABIN: [*Sound of aircraft-overspeed warning*]
CAPTAIN: Okay, it's no matter. Pull the airspeed [back]. We will see. . . .
CABIN: [*Overspeed warning stops*]
COPILOT: Now it is three hundred and fifty [mph], yes.
CAPTAIN: Let's take that like this. . . .
CABIN: [*Sound of four warning-alert tones; sound of stick shaker starts, indicating a stalling aircraft, and continues to end of recording; sound of four warning-alert tones*]
COPILOT: Sir . . . [put the] nose down.
FLIGHT ENGINEER: Now . . .
COPILOT: [Increase] thrust. . . .
CAPTAIN: Disconnect the autopilot. Is autopilot disconnected?
COPILOT: Already disconnected, disconnected, sir.
CAPTAIN: [It will] not climb? What am I to do?
COPILOT: You may level off. Altitude okay. I am selecting the altitude hold, sir.
CAPTAIN: Select, select.
COPILOT: Altitude hold. Okay, five thousand feet.
CAPTAIN: Thrust levers, thrust, thrust, thrust, thrust.
COPILOT: Retard.
CAPTAIN: Thrust. Don't pull back, don't pull back, don't pull back, don't pull back.
COPILOT: Okay, open [the throttles], open. . . .
CAPTAIN: Don't pull back. Please don't pull back.
COPILOT: Open, sir, open. . . .
FLIGHT ENGINEER: Sir, pull up.
CAPTAIN: What's happening?
COPILOT: Oh, what's happening?

CABIN: [*Ground-proximity warning system alarm, "Whoop, whoop, pull up," continues until the end of the recording*]
COPILOT: Let's do like this. . . .

END OF TAPE

The Birgen Air charter stalled and went into free fall for one minute and forty-one seconds before hitting the sea. All aboard were lost.

Elmendorf Air Force Base, Alaska

September 22, 1995

■

Yukla 27

A Boeing 707 configured as a radar E-3A for the U.S. Air Force's 3rd Wing prepared for takeoff at 7:47 A.M. on September 22, 1995, with a crew of four and twenty passengers. The morning was bright but overcast. Because of the fall season, flights of migrating Canada geese had landed on the grasslands around the airfield, while other geese had taken off and were flying in the vicinity of the runway. The crew of the air force Boeing, called Yukla 27 heavy, was aware of the birds.

We pick up the CVR just as Yukla 27 prepared for takeoff.

ELMENDORF TOWER: Yukla Two seven heavy, the wind three one zero at one one, cleared for takeoff [on] Runway Five. Traffic is a C–One thirty three miles north of Elmendorf northwest-bound, climbing out of two thousand.

CABIN: [*Light switches*]

COPILOT: [The C-130] in sight. And Yukla Two seven heavy cleared for takeoff, traffic in sight. [*To cockpit*] Cleared for takeoff, crew.

FLIGHT ENGINEER: Check complete.

CABIN: [*Engines spool up*]

CAPTAIN: Engineer, set takeoff power.

CABIN: [*Engines spool up*]

CAPTAIN: Eighty knots, copilot's aircraft [copilot is flying the airplane]. Your airplane. Vee one. Rotate.

COPILOT: [Look at] all the birds.

FLIGHT ENGINEER: [There are a] lotta birds here.

At this point, the number-one and -two engines ingested several geese, disintegrating the engine's fan blades.

CAPTAIN: Damn, we took one.
COPILOT: What do I got?
FLIGHT ENGINEER: We took two of 'em.
CAPTAIN: We got two motors.
FLIGHT ENGINEER: Flight start.

The flight engineer calls for an attempt to restart the failed engines.

COPILOT: Roger that.
CAPTAIN: Take me to override.
COPILOT: Go to override on, on the . . . Elmendorf Tower, Yukla Two seven heavy has an emergency. Lost, ah, number-two engine. We've taken some birds.
AN INSTRUCTOR FLIGHT ENGINEER IN COCKPIT: You're in override. There's the rudder.
FLIGHT ENGINEER: Got it.
INSTRUCTOR: You're in override.
CAPTAIN: Thank you.
FLIGHT ENGINEER: Starting [to] dump fuel.
CAPTAIN: Start dumping.
TOWER: Yukla Two seven heavy, roger. Say intentions.
CABIN: [*Stick shaker activates, continues until impact*]
COPILOT: Yukla Two seven heavy's coming back around for an emergency return. Lower the nose, lower the nose, lower the nose.
TOWER: Two seven heavy, roger.
CAPTAIN: Goin' down.
COPILOT: Oh my god.
CAPTAIN: Oh, shit.
COPILOT: Okay, give it all you got, give it all you got. Two seven heavy, emergency . . .
TOWER: Roll the crash [equipment], roll the crash—
COPILOT: [*On public-address system*] Crash [landing].
CAPTAIN: We're goin' in. We're goin' down.

END OF TAPE

The air force concluded that the cause of the crash was a "multiple birdstrike" of Canada geese in the number-one and -two engines just after takeoff. The aircraft crashed into woods and caught fire. All twenty-four on board were lost.

Amsterdam, Netherlands

October 4, 1992

■

El Al Flight 1862

El Al Flight 1862, with a crew of three and one passenger, was ferrying freight from Amsterdam's Schiphol Airport to Ben Gurion Airport in Israel. The Boeing 747 took off from Amsterdam Runway 01L at 5:21 in the afternoon. Six minutes later, at 5:27:30, as the aircraft was climbing through 6,500 feet, the number-three engine and the pylon that attaches the engine to the wing separated from the wing and slammed into the number-four engine, which also separated from the wing. The crew immediately declared an emergency and requested Runway 27 for landing. There is no identification on the CVR tape's transcript of the individual voices. "Crew" is used instead of the individuals in the cockpit.

We pick up the CVR just as the crew is declaring an emergency.

CREW: El Al One eight six two, Mayday, Mayday, we have an emergency. . . .

Air-Traffic Control acknowledges the declaration of an emergency and immediately warns all other aircraft in the vicinity away from the airport with the term "break."

AIR-TRAFFIC CONTROL: El Al One eight six two, roger. Break, KLM Two three seven, turn left heading zero nine zero. El Al One eight six two, do you wish to return to Schiphol?

CREW: Affirmative. Mayday, Mayday, Mayday.

CONTROL: Turn right heading two six zero, field, eh . . . behind you, eh . . . in your—to the west, eh . . . distance eighteen miles.

CREW: Roger, we have fire on engine number three, we have fire on engine number three.

CONTROL: Roger, heading two seven zero for downwind.

CREW: Two seven zero downwind.

CONTROL: El Al One eight six two, surface wind zero four zero at twenty-one knots.

CREW: Roger.

CREW: El Al One eight six two, lost number-three and number-four engines, number-three and number-four engines.

CONTROL: Roger, One eight six two.

CREW: What will be the runway in use for me at Amsterdam?

CONTROL: Runway Six in use, sir. Surface wind zero four zero at twenty-one knots. . . .

CREW: We request [Runway] Twenty-seven for landing.

CONTROL: Roger, can you call Approach now, [tune to radio frequency] one two one point two for your lineup?

CREW: One two one point two, bye-bye.

CONTROL: Bye.

CREW: Schiphol, El Al One eight six two, we have an emergency, eh . . . we're . . . number-three and -four engines inoperative . . . [returning] landing.

CONTROL: El Al One eight six two, roger, copied [understood] about your emergency, contact [radio frequency] one one eight point four for your lineup.

CREW: One one eight point four, bye. Schiphol, El Al One eight six two, we have an emergency, number-three and number-four engines inoperative, request [Runway] Twenty-seven for landing.

CONTROL: You request Twenty-seven, in that case heading three six zero . . . three six zero the heading. Descend to two thousand feet on one zero one two, mind, the wind is zero five zero at twenty-two.

CREW: Roger, can you say again the wind, please?

CONTROL: Zero five zero at twenty-two.

CREW: Roger, what heading for Runway Twenty-seven?

CONTROL: Heading three six zero, heading three six zero and [then] give you a right turn on, to cross the localizer first, and you've got only seven miles to go from [your] present position.

CREW: Roger, three six [heading 360] copied.

CONTROL: El Al One eight six two, what is the distance you need to touch down?

CREW: Twelve miles final we need for landing.

CONTROL: Yeah, how many miles [for] final [approach] . . . eh, correction . . . how many miles track miles you need?

CREW: We need . . . eh . . . a twelve miles final for landing.

CONTROL: Okay, right right heading one zero zero, right right heading one zero zero.

CREW: Heading one zero zero.

CONTROL: El Al One eight six two, just to be sure, your engines number three and four are out?

CREW: Number three and four are out and we have . . . eh . . . problems with our flaps.

CONTROL: Problem with the flaps, roger.

CREW: Heading one zero zero, El Al One eight six two.

CONTROL: Thank you, One eight six two.

CREW: Okay, heading . . . eh . . . and turning, eh . . . maintaining . . .

CONTROL: Roger, One eight six two, your speed is?

CREW: Say again?

CONTROL: Your speed?

CREW: Our speed is . . . eh . . . two six zero.

CONTROL: Okay, you have around thirteen miles to go to touch down. Speed is all yours. You are cleared to land Runway Twenty-seven.

CREW: Cleared to land Twenty-seven.

CONTROL: El Al One eight six two, a right right turn heading two seven zero. Adjust on the localizer. Cleared for approach.

CREW: Right right two seven zero.

CONTROL: El Al One eight six two, you're about to cross the localizer due to your speed. Continue the right turn heading two nine zero, heading two nine zero, twelve track miles to go, twelve track miles to go.

CREW: Roger, two nine zero.

CONTROL: El Al One eight six two, further right, heading three one zero . . . heading three one zero.

CREW: Three one zero.

CONTROL: El Al One eight six two, continue descent fifteen hundred feet . . . fifteen hundred.

CREW: Fifteen hundred, and we have a controlling problem.

CONTROL: You have a controlling problem as well, roger.

CREW: Going down, One eight six two, going down . . . going down. . . .
CONTROL: Copied, [you are] going down.
CREW: Raise all the flaps, all the flaps raise, lower the gear.
CONTROL: Yes, El Al One eight six two, your heading—

END OF TAPE

Metal fatigue was discovered in the inboard midspar of the number-three engine pylon. This caused the pylon and engine to separate from the wing. Part of the leading edge of the wing was damaged by the impact with the number-three and -four engines, and, as a result, the use of several control systems was lost or limited. The investigator concluded that the flight crew had such limited control of the airplane that a safe landing was highly improbable, if not virtually impossible.

Flight 1862 crashed into an apartment complex, killing the crew of three and the single passenger. Forty-seven people on the ground died.

Nagoya-Komaki International Airport, Japan

April 26, 1994

■

China Airlines Flight 140

China Airlines Flight 140, an Airbus, took off from Taipei, Taiwan, for a flight to Nagoya, Japan, where it arrived over the outer marker at 8:12:26 P.M. The copilot started an approach to Runway 34 ILS with auto-throttles engaged at an altitude of 1,070 feet when he inadvertently and unknowingly pushed the Take Off–Go Around (TOGA) switch on the control panel, which automatically powered up the engines and configured the airplane for an aborted landing and "go-around" of the airport. Unaware of the instructions they had given to the autopilot, the crew fought the airplane to make it land. To remain on the glide slope, the copilot disengaged the auto-throttle, reduced thrust manually, and switched off the autopilot.

At 1,030 feet the crew commanded the autopilot to put the plane on the glide slope. But the autopilot reverted to the go-around mode. To get back on the glide slope, the crew lowered the aircraft while the autopilot was telling the airplane to pitch up and go around. The captain and copilot were fighting the autopilots for control of the aircraft.

Forty-two seconds after the mistaken go-around selection, the crew disengaged the autopilots again, but the aircraft kept climbing. Eight seconds afterward, the computer, which had been told to go around, triggered maximum thrust of the Airbus engines. The captain pulled back on the engine thrust and the speed dropped. The Airbus stalled at 1,800 feet.The aircraft hit the ground tail-first 300 feet to the right of the runway and burst into flames.

The weather that evening was clear with light winds and nine miles of visibility.

The Airbus was carrying a crew of 15 and 271 passengers.

We pick up the CVR tape with Flight 140 cleared for landing and on the glide slope. The copilot is at the controls.

CAPTAIN: Engage it. Push it. It's too high. You . . . on go-around mode. Don't worry, slowly, slowly, begin it. Support it firmly with your hand. Push, push it.
COPILOT: It could not be pushed.
CAPTAIN: Don't worry, do it slowly. Okay, I try.
COPILOT: I engage it. I engage it.
CAPTAIN: What is this?
COPILOT: I engage it.
CAPTAIN: God damn it! Why it comes in this way?

Flight 140 had aborted its landing approach, and the automatic go-around system was pitching up the nose.

NAGOYA-KOMAKI TOWER: Stand by further instruction.
CREW: Aircraft will stall at this rate.
COPILOT: No way! No way!
CAPTAIN: Set, set, set it. Don't worry. Don't worry. Don't upset. Don't upset.
CABIN: [*A recorded warning voice says, "Terrain, terrain."*]
CREW: Power!
CABIN: [*Stall warning says, "Pull up, pull up."*]
COPILOT: Ah . . . no way! No way!

END OF TAPE

The crew was confused about the autopilot and auto-throttle. The crew might have saved the aircraft even in the final seconds before it crashed if they had reverted to basic flight procedures—indeed, if they had switched off the autopilot and flown the airplane with their own hands.

The crew of 15 died in the resulting fiery crash; of the 271 passengers aboard, all but 7 perished.

Cheju Island Airport, South Korea

August 10, 1994

■

Korean Air Flight 2033

Korean Air Flight 2033, an Airbus A300, was approaching the rain-slicked airport that summer morning, when a conflict arose between the captain of the airplane and his copilot.

The copilot, Chung Chan Kuy, a Korean, asked Captain Barry Woods, a Canadian, several times whether he wanted to "go around." Apparently, the copilot was concerned about the length of Runway 06. When the captain told his copilot not to go around, the copilot grabbed for the throttles. Woods told him to "get off!"

They were flying through a tropical storm with winds that were gusting up to 30 mph.

We pick up the CVR tape just as the runway comes into sight.

COPILOT: Runway in sight, runway in sight.
CAPTAIN: I got it, I got it.
COPILOT: Okay. Right side? Right?
CAPTAIN: Yeah . . . okay. Give me the, uh, four hundred [feet] . . . three. . . . Minimum sink rate.
COPILOT: Sink rate, sink rate.
CAPTAIN: Okay, okay.
COPILOT: One hundred [feet]. Speed?
CAPTAIN: Yeah, fifty [feet].

At this juncture, the copilot decides that there is insufficient distance for the Airbus to land safely without crossing the end of the runway. The copilot, when the aircraft is only thirty feet off the ground, decides to go around and pulls back on the yoke.

COPILOT: Go around, forty [feet], thirty [feet]. . . .
CAPTAIN: Get your hands . . . get off! Get off! Tell me what [the altitude] is. Twenty [feet]. Get off.
COPILOT: Go around?
CAPTAIN: No, no, ten, five [feet].

Here they touch down and brakes and the thrust reversers are deployed; no matter, the copilot still wants to abort the landing and go around against the wishes of the captain, who is telling him not to go around but to brake the aircraft.

CAPTAIN: Reverse [engines]. Brake. What are you doing? Don't. What, man . . . ? You're gonna kill us. Hold yoke.
CABIN: [*Sound of crash*]

The aircraft crashes through a fence and slides into a rice paddy on the side of the field, coming to a stop.

CAPTAIN: Yup, okay. You all right?
COPILOT: All right.
CAPTAIN: Okay. Get this off. What did you pull us off [for]? What did you pull us off [for]?
COPILOT: [I wanted to] go around . . . go around.
CAPTAIN: Yeah, but we were on . . . we were on the runway. Why did you pull us off? Okay, okay. We got to get out of here. Open your window.
CABIN: [*Sound of the cockpit window opening*]
CAPTAIN: Get your [evacuation] slide. Why did you pull us off? We had full reverse on. Pull the fire handles. Pull 'em.
COPILOT: Fire pulls.
CAPTAIN: Okay, get out. Get out.

END OF TAPE

Miraculously, in spite of the airplane catching fire, the 6-member cabin crew got all of the 152 passengers off the airplane safely, with only a few minor injuries. The pilot and copilot exited through the cockpit windows. The Airbus was a total write-off. Korean authorities contemplated bringing criminal charges against the captain and copilot of the aircraft.

Tokyo-Haneda Airport, Japan

August 12, 1985

■

Japan Airlines Flight 123

JAL Flight 123, a Boeing 747, took off from Tokyo-Haneda at 6:12 P.M. for a flight to Osaka with 509 passengers and 15 crew members. At 6:24 P.M., while climbing through 23,900 feet at a speed of 345 mph, an unusual vibration and the sound of an impact signaled a catastrophic rupture of the aft bulkhead.

In the two minutes that followed the rupture, hydraulic pressure dropped. Ailerons, elevators, and yaw damper—surfaces that control the direction and altitude of the aircraft—failed to work. The plane was out of control as it fell to 6,600 feet. The crew tried to use engine thrust to maneuver the aircraft. At 6,600 feet, the airspeed dropped dangerously to 124 mph. Then the aircraft climbed sharply to 13,400 feet and started to fall again. It finally brushed against a tree-covered ridge along a mountain but continued to fly for another three seconds, until it struck another nearby ridge and burst into flames. It was the worst disaster in Japanese aviation history.

We begin the CVR tape just after the collapse of the aft bulkhead.

CABIN: [*Public-address system: "Put on oxygen mask. Fasten seat belt. We are making an emergency descent."*]

FLIGHT ATTENDANT: [*On public-address system to the passengers*] May we ask you passengers with infants . . . nearby passengers . . . please be ready.

CAPTAIN: Right turn. Right turn.

COPILOT: I did it.

CAPTAIN: Yes. Do not bank so steeply.

COPILOT: Yes, sir.
CAPTAIN: Do not bank so steeply.
COPILOT: Yes, sir.
CAPTAIN: Recover it.
COPILOT: It does not recover.
CAPTAIN: Pull up.

Thirteen seconds go by while the crew tries to gain control of the aircraft.

CAPTAIN: All hydraulics failed.
COPILOT: Yes, sir.
CAPTAIN: Descend.
COPILOT: Yes, sir.

Another fifteen seconds go by.

CAPTAIN: Hydro [hydraulic] pressure is lost.
COPILOT: All lost?
CAPTAIN: All lost.

Two and a half minutes go by.

FLIGHT ENGINEER: How about cabin pressure? Have cabin masks dropped? So? Well, cabin pressure, please.

Slightly more than a minute passes.

FLIGHT ENGINEER: Cabin pressure dropped. What? More aft . . . ah. What was damaged? Where? Coat room? Ah, coat room . . . general . . . It dropped in baggage space. It would be better to land. All persons inhale [into] masks. Captain?
CAPTAIN: Yes?
FLIGHT ENGINEER: Do we put on [our] masks?
CAPTAIN: Yes, it would be better.
FLIGHT ENGINEER: I think it would be better to inhale emergency mask.
CAPTAIN: Yes.

For another three minutes the Boeing 747 flounders, with the pilot using the engines to "fly" the airplane by applying power to the engines to steer the aircraft left or right.

CAPTAIN: Nose down. Lower nose.
COPILOT: Yes, sir.
CAPTAIN: Lower nose.
COPILOT: Yes, sir.
CAPTAIN: Lower nose.
COPILOT: Yes, sir.
CAPTAIN: Stop saying that. Do it with both hands, with both hands.
COPILOT: Yes, sir.
FLIGHT ENGINEER: Do we [put the landing] gear down?
COPILOT: Gear down?
CAPTAIN: Speed does not drop.
COPILOT: Speed does not drop.
CAPTAIN: Lower nose.
COPILOT: Yes, sir.

For the next ten minutes, the crew wrestles with the controls and the throttles, from applying power to lowering the nose.

CAPTAIN: Power, power, lower nose. It is heavy. Lower nose slightly more. It is heavy. It is heavy. Oh, it is heavy.
COPILOT: What about flap? Do I extend it?
CAPTAIN: It is a bit too early.
COPILOT: It is too early?
CAPTAIN: It is too early, too early. Was gear extended?
COPILOT: Gear was extended [lowered].
CAPTAIN: Lower nose.
FLIGHT ATTENDANT: [*On public-address system to the passengers*] Those passengers who accompany babies, please keep your heads on seat backs and hold your babies firmly. Did you fasten seat belts? All tables are retracted. Please check it. . . . Sometimes we land without notice. . . . We keep contact with ground. . . .
CAPTAIN: Lower nose. It may be hopeless. It is a mountain? Raise nose.
COPILOT: Yes, sir.
CAPTAIN: Control to right. Right turn. Raise nose. We may hit the mountain.
COPILOT: Yes, sir.
CAPTAIN: Right turn.
CABIN: [*Sound of ground-proximity warning system*]
CAPTAIN: Max power.

COPILOT: Max power.
CAPTAIN: Hold out.
CAPTAIN: Left turn.
COPILOT: Yes, sir.
FLIGHT ENGINEER: Hold out.
CAPTAIN: Left turn, now . . . Left turn.
COPILOT: Decrease power slightly.
CAPTAIN: Right, right . . . Lower nose.
COPILOT: Fully controlled now.
CABIN: [*Warning sounds stop*]
COPILOT: It is totally ineffective. . . .
CAPTAIN: Lower nose.
COPILOT: Good.
CAPTAIN: Here we go.
COPILOT: Yes, sir.
FLIGHT ENGINEER: Do I increase power?
CAPTAIN: Power, power.
CABIN: [*Warning sounds*]
CAPTAIN: Power. Let's increase power, power. . . .
CABIN: [*Stall warning sounds: "Pull up, pull up."*]
CAPTAIN: Ah, it is hopeless. Max power.
COPILOT OR FLIGHT ENGINEER: Max power.
CAPTAIN: Stall. Let's hang tough. Hold out.
COPILOT: Yes, sir.
CAPTAIN: Lower nose. Hold out, hold out.
COPILOT: Fully controlled.
FLIGHT ENGINEER: Speed is controlled by power. How about power control? Power control, Captain?
CAPTAIN: Yes. Speed is two twenty knots.
FLIGHT ENGINEER: Yes, sir.
CAPTAIN: Do not lower nose. It increases speed.
COPILOT: Yes, sir.
CAPTAIN: It's going down.
COPILOT: Yes, sir. Raise nose, raise.

Five minutes go by, while the crew struggles with the controls.

CAPTAIN: Raise nose, raise nose, raise nose. Stop flap. Do not extend flap so deeply. Flap up, flap up, flap up, flap up, flap up.
COPILOT: Yes, sir.

CAPTAIN: Power, power. Flap.
COPILOT: I am retracting [the flaps].
CAPTAIN: Raise nose, raise nose, power.
CABIN: [*Ground-proximity warning: "Sink rate. Pull up, pull up, pull up, pull up."*]
CABIN: [*Sound of collision with the first peak*]

Three seconds go by.

CABIN: [*Sound of collision with the second peak*]

Two seconds go by.

END OF TAPE

Metal-fatigue cracks caused the collapse of the aft bulkhead, which broke through the hydraulic controls that extend along the fuselage to the cockpit of the Boeing 747. Five hundred five people died that day. Four passengers miraculously survived. Although none of the information appeared on the CVR tape, the passengers were aware of their peril. After the crash, several last letters from passengers to loved ones were discovered in the wreckage, saying good-bye.

Mulhouse-Habsheim, France

June 26, 1988

■

Air France Flight 296Q

The cockpit crew was about to show off the new Air France Airbus A320 that spring day to crowds waiting below at the Habsheim Airfield. Demonstrating the new airplane's capabilities to those on the ground and to 130 passengers, mostly members of the press, who were riding along, the crew planned two passes of the airfield, one a low-level, high-speed "flyby." For everyone that day, it would be a demonstration from hell.

The crew started out in a jovial mood, laughing at the antiquity of an airplane that had landed and crossed their path as they taxied the Airbus out for takeoff. A flight attendant chatted as the crew went through its preflight checklist.

We pick up the CVR as the copilot asks the captain to explain what they are going to do for the demonstration.

CAPTAIN: Okay, then. [We're going to] take off, [then make a] right turn. Leave the flap at one. Anyway, we do normal takeoff, retract the landing gear, and with flaps at one, we go nice and easy. As soon as we have formally identified [the airport], we extend the flaps to three, landing gear extended, [and we] do the flyover at one hundred feet, landing gear out, and then you leave it to me. . . .

CAPTAIN: I've done it twenty times, that one.

COPILOT: Okay, we're agreed.

CAPTAIN: And then, after, we bring everything [flaps] in, move off, and give it all it's got to three forty knots [391 mph], and [on the] second [flyover], you also go over at one hundred feet. And there's no need to pull two point five g's, as back there [in the passenger cabin], they won't like it.

COPILOT: Okay, all that.

TOWER: AF Two nine six Q, Basle?

COPILOT: Go ahead.

TOWER: Yes, Two nine six Q, can you tell me what altitude you want?

COPILOT: Two thousand [feet].

CAPTAIN: Roger, one thousand feet to get there, one thousand feet to get there, one thousand feet go around . . .

COPILOT: Yes, after takeoff, we'll take a right turn one thousand feet, then there we'll make two flyovers, the second [one] at high speed.

TOWER: Roger, Two nine six Q.

CAPTAIN: So it's one thousand feet Fox Echo.

COPILOT: [*To Tower*] Well, it would be good if we could have clearance before the second flyover as we'll be passing at high speed, like that. We could climb quickly.

TOWER: Roger, Two nine six Q. You confirm one thousand feet above ground then?

COPILOT: One thousand feet above ground, and by right turn there. Then we'll descend for low-altitude flyover of the airfield there.

TOWER: Okay, I'll call you.

Still on the taxiway, the crew goes over the plans for the flyover several times. They inform the Tower of what they are planning, and they ask for clearances. However, the copilot still isn't clear about what he is supposed to do.

COPILOT: Between the two [flyovers], when I take it for the high-speed flyover, you negotiate the one-ninety-degree clearance so that we can get out of there and do a pull-out at two g's.

CAPTAIN: No, don't pull out at two point five g's, as yesterday we did two g's. Two point five g's in training, and I've had enough of that, I have. . . .

PURSER: [*Coming into cockpit*] In short, what are we doing?

CAPTAIN: We take off, fly to the Habsheim Airfield. There we do a low-altitude flyover, bank, accelerate, fly over at top speed, and get out of there.

COPILOT: [*To purser*] Yes? You were saying?
PURSER: No, but I was saying [that] the cabin is ready.
COPILOT: You stay seated during the complete flyover.
PURSER: About the announcements. You're doing something?
CAPTAIN: I'm going to tell them [the passengers], yes. The announcements, I'll do them.
PURSER: Okay. There are also some [Germans on board].
CAPTAIN: [*Over public-address system to the passengers*] Ladies and gentlemen, hello and welcome aboard this Airbus Three twenty, number three of the series for Air France and which has only been in service for two days. We shall take off for a short tourist flight starting at the Habsheim flying club, where we will do two flyovers to demonstrate the [quality] of French aviation. . . .
TOWER: Hello, AF Two nine six Q. Hold before Runway Sixteen.

For the next two minutes, the crew goes through the pre-takeoff checklist. The two men discuss the computer flight-control inputs. The flight attendant comes into the cockpit.

CAPTAIN: [*To flight attendant*] What? No. I'll tell you something. You're not used to cockpits, are you? But like this, when we're really busy, as you can see, you should not speak, ah, shouldn't bother to speak, shouldn't bother the pilots. Eh? You see how it is?

The captain and the copilot converse with two passengers who come up to the cockpit. They discuss the possibility of flying over another destination, Meribel, for which they do not have clearance, after the flyover of Habsheim.

TOWER: AF Two nine six Q, clear for takeoff Runway Sixteen, wind calm.
CAPTAIN: Take off. Go.
CABIN: [*Sound of engine acceleration*]
COPILOT: Parameters normal.
CAPTAIN: *Gut* [good]. One hundred [mph].

The copilot seems amazed at the flight characteristics of the new aircraft.

COPILOT: It [the computer] is requesting "climb," the bastard. You see that?
CAPTAIN: Yeah, that happens.
COPILOT: Yeah, yeah. Vee one, Vee rotate . . .
CAPTAIN: Postive climb. Gear up.
COPILOT: Gear up.
CAPTAIN: Gear up?
CABIN: [*Reduction in engine noise*]
COPILOT: Gear up, flaps one, after-takeoff checklist is completed, except for flaps.
CAPTAIN: I'm disengaging the auto-throttle.
CABIN: [*Repetitive gongs*]
CAPTAIN: That's the landing gear, that. It's not important. Kill it, eh [the gong]? Oh, kill it. It gets on my nerves, that one.
COPILOT: You're eight nautical miles there. You'll soon see it [Habsheim Airfield]. There's the motorway.
CAPTAIN: We'll leave the motorway to the left, won't we . . . ? It's to the left. . . . No, to the right of the motorway.
COPILOT: It's slightly to the right of the motorway. So you, you leave the motorway on the left.
CAPTAIN: Okay. As soon as we identify [Habsheim Airfield], we [will] descend very quickly then. . . .
COPILOT: Leave everybody fastened in [their seat belts].
TOWER: Roger, AF Two nine six. You can contact Habsheim one two five point two five. Good-bye.
COPILOT: [*To Tower*] One two five point two five, good-bye. Habzeim? [*To captain*] That's it, no?
CAPTAIN: [*Correcting his pronunciation*] Habsheim. Habs . . . heim . . .
COPILOT: Habsheim. Habsheim Air Charter, Two nine six Q. Hello.
CAPTAIN: There's the airfield. It's there.You got it, have you?
TOWER: Two nine six Q, hello.
CAPTAIN: What? We're coming into view of the airfield for flyover.
TOWER: Yes, I can see you. You're cleared, eh. Sky is clear.
CAPTAIN: Gear down.
COPILOT: [*To Tower*] Okay, we're going in for the low-altitude, low-speed flyover, Two nine six Q.
TOWER: Roger.
CAPTAIN: Flaps two.
COPILOT: Good. Gear is down. Flaps two.

CAPTAIN: Flaps three.
COPILOT: Flaps three.
CAPTAIN: That's the airfield. You confirm?
COPILOT: Affirmative.
CABIN: [*Gongs sound to warn of too low terrain, two hundred feet*]
COPILOT: Okay, you're at a hundred feet. Watch, watch. . . .
CABIN: [*Mechanical voice calls out the altitude in feet: "One hundred," then "Fifty . . . forty . . . thirty . . . "*]
CAPTAIN: Okay, I'm okay there. Disconnect auto-throttle.
COPILOT: Watch out for the pylons ahead. See 'em?
CAPTAIN: Yeah, yeah. Don't worry.
COPILOT: Go around track.
CABIN: [*Increase in engine speed; noises of impact*]
CAPTAIN: Shit.

END OF TAPE

The crew fought the autopilot and finally lost control of the Airbus, which could not regain enough altitude to clear the trees at the end of Habsheim Airfield. Three passengers died in the resulting crash. A combination of errors, including pilot errors and confusion over the operation of the Airbus's computer command control, caused the crash. This tragedy was a dreadful embarrassment to French aviation.

15

Dallas–Fort Worth, Texas

August 2, 1985

■

Delta Air Lines Flight 191

A Delta Air Lines Lockheed L-1011 TriStar, Flight 191, was on final approach into Dallas–Fort Worth, Texas, from Fort Lauderdale–Hollywood, Florida, when it encountered heavy thunderstorms. While approaching Runway 17L, the crew could see flashes of lightning in a cumulonimbus cloud. The TriStar, with a crew of 11 and 152 passengers, stayed on the approach despite the rain and wind. At 6,300 feet from the end of the runway, the TriStar entered a low-level microburst, which created massive wind shear. At that time, the crew lacked definitive, real-time training to learn to avoid and escape low-level wind shear. Caught by surprise, the crew could only sit helplessly. As with other crashes, the double tragedy of Delta 191 was the ignorance of the crew. They simply did not understand the nature of wind shear.

We start the CVR at the moment when the Tower asks another Delta jet—Flight 963—to change direction. That crew demurs, because it sights the thunderstorms ahead. The cockpit crew of Delta 191 hears the exchange.

DALLAS–FORT WORTH TOWER: Delta Nine six three, I got an area twelve miles wide [that] all the aircraft are going through. All the aircraft are going through there. [It's a] good ride. I'll have [you] turn back in before you get to the weather [thunderstorms].

Now the captain and copilot of Delta 191 can be heard on the cockpit tape.

COPILOT: It would be nice if we could deviate to the south of two five zero.

CAPTAIN: Somebody [Delta Flight 963] just ahead of us tried and they wouldn't let them do it. They're working a twelve-mile corridor. The airplanes that have been going through there have been all right.

TOWER: Delta One nine one, descend and maintain one zero thousand [10,000 feet], altimeter two niner niner one.

FLIGHT ENGINEER: Think that might have been for us, guys.

CAPTAIN: [*To Tower*] Sorry, was that for Delta One nine one?

TOWER: One nine one, descend and maintain one zero thousand, the altimeter two niner niner one, and suggest now a heading of two five zero. . . . We have a good area there to go through.

CAPTAIN: Well, I'm looking at a [microburst] cell at about heading, ah, two five five, and it's a pretty good-size cell, and I'd rather not go through it. I'd rather go around it, one way or another.

TOWER: I can't take you south. I got a line of departures to the south. I've had about sixty aircraft go through this area out here, ten . . . twelve miles wide. They're getting a good ride. No problems.

CAPTAIN: Well, I see a cell now about heading two four zero.

TOWER: Okay . . . when I can I'll turn you. It'll be about the zero one zero radial.

COPILOT: [*To captain*] He [the Tower] must be going to turn us before we get to that area [of thunderstorms and cells].

CAPTAIN: Put the girls down [tell the flight attendants to be seated].

CABIN: [*Sound of chime*]

About three minutes later, the crew questions the experience of the controller in the tower.

CAPTAIN: Getting kind of hot in the oven with this controller. See? That's what the lack of experience does. You're in good shape. I'm glad we didn't have to go through that mess. I thought sure [the controller] was going to send us through it.

FLIGHT ENGINEER: Looks like it's raining over Fort Worth.

COPILOT: Yeah.

Five minutes go by. The crew begins its final approach check.

TOWER: Attention, all aircraft listening . . . There's a little rain shower just north of the airport and they're starting to make ILS approaches. . . .

Delta 191 descends through 7,000 to 5,000 feet. After completing more pre-landing checks, the pilot sees the storm out the window.

TOWER: Delta One nine one, turn left heading one nine zero, and I'll turn you right back on [downwind] in just a second.
CREW: All that screwing around for nothing.
COPILOT: We're going to get our airplane washed [with rain].
CAPTAIN: What?
COPILOT: We're going to get our airplane washed.

For the next three minutes, Delta 191 lines up for approach and landing.

TOWER: And we're getting some variable winds out there due to a shower on short out there, north end of DFW [Dallas–Fort Worth].
CREW [unidentified, either captain or copilot]: Stuff [thunderstorm] is moving in.
CAPTAIN: One six zero is the speed.
CAPTAIN: Tower, Delta One nine one, out here in the rain. Feels good.
FLIGHT ENGINEER: Landing gear [down].
TOWER: Delta One ninety-one heavy, One seven left, cleared to land, winds zero nine zero at five gusting to one five.
COPILOT: All right. Landing gear?
CAPTAIN: Down, three green [confirmation that gear is down].
FLIGHT ENGINEER: Flaps, slats.
COPILOT: Straight through eighteen [1,800 feet].
CABIN: [*Sound of altitude-alert horn*]
COPILOT: There's lightning coming out of that one [cloud].
CAPTAIN: What?
COPILOT: There's lightning coming out of that [cloud].
CAPTAIN: Where?
COPILOT: Right ahead of us. We're too late.

TOWER: [*To a Delta flight on the taxiway*] Delta Ten sixty-one, cross One seven right without delay, ground point six five after you cross.

DELTA 1061: Say again, Ten sixty-one.

TOWER: Ten sixty-one, cross One seven right, ground point six five.

CAPTAIN: How about the DME [Distance-Measuring Equipment]?

COPILOT: Well, you haven't had it for the last five minutes.

TOWER: [*To other taxiing airplanes*] Delta Nine sixty-three and American Six nineteen, cross One seven right, ground point six five, after you cross.

Up to this point in the flight, neither the passengers nor the crew have a hint of what is to come; as far as the passengers are concerned, their flight is nearly over. The crew is probably feeling some anxiety about the weather, but because other flights have landed safely only minutes in front of them, they do not appear to be overly worried.

FLIGHT ENGINEER: A thousand [feet].

COPILOT: One thousand feet.

CAPTAIN: Watch your speed.

TOWER: [*To a taxiing aircraft*] American One six six, contact Departure.

Now the Delta TriStar enters the microburst and the wind shear.

CAPTAIN: You're gonna lose it all of a sudden. There it is. Push it up, push it way up!

COPILOT: [Pull] way up.

CAPTAIN: Way up. Way up.

CABIN: [*Sound of engine, high rpm*]

CAPTAIN: That's it. Hang on to the son of a bitch.

COPILOT: What's the Vee ref?

TOWER: [*To a taxiing aircraft*] American Five eighty-six, taxi into position and hold One seven right.

AMERICAN 586: Position and hold.

CABIN: [*Ground-proximity warning, mechanical: "Whoop, whoop! Pull up!"*]

CAPTAIN: TOGA [go around, abort landing approach]!

CABIN: [*"Whoop, whoop! Pull up!" Sound of go-around initiation; beeps*]
CAPTAIN: Push it way up!

END OF TAPE

Delta 191 hit the ground 6,300 feet short of the runway. It then slid along the ground and struck a car on a highway, collided with two water tanks, broke up, and burst into flames. Eight of the 11 crew died. Twenty-six of the 152 passengers survived the crash.

16

Kelly Air Force Base, Texas

October 4, 1986

■

Southern Air Transport Flight (Logair) 15

The events surrounding the crash of Southern Air Transport Flight 15 unfolded a little like a mystery.

The Lockheed L-382G four-propeller cargo aircraft arrived in Texas in the early-morning hours, with a crew of three and a cargo of Class B and Class C-3 explosives (dynamite propellants and rocket motors for fighter-jet ejection seats). Southern Air was a domestic airline that hauled cargo for the military. Flight 15, known as Logair 15, had originated at Hill Air Force Base, Utah. The new crew had arrived early in the morning of October 4 at Kelly Air Force Base Flight Operations, where they spoke briefly with the retiring crew, just in from Utah. The retiring captain said the Lockheed was "in good shape."

Flight 15's flight engineer supervised the loading and unloading of cargo to make certain that each of the ten pallets of cargo was secured both forward and aft on the airplane's floor locks. The military loading supervisor recalled later that the airplane's elevator—the control surface on the tail that flies the airplane up or down—was raised to prevent its being damaged during the loading operations.

By four that morning, the crew of Flight 15 was ready to leave. It went through all the preflight checks before taxiing out to Runway 15. At 4:05 A.M., the captain notified the Tower that they were waiting for permission to take off, on a Visual Flight Rules ascent.

We pick up the CVR tape here.

TOWER: Logair One five, one thousand feet closed wind one five zero at four—cleared for takeoff.

COPILOT: One five, roger. Transponders on.
CAPTAIN: Okay, before-takeoff check . . .
FLIGHT ENGINEER: Okay, whenever you're ready for [engines] one and four, Captain.
CAPTAIN: One and four normal.
COPILOT: Off at zero five [4:05 in the morning].
CAPTAIN: Yeah.
COPILOT: Final.
FLIGHT ENGINEER: Before-takeoff check is complete.
COPILOT: Finals clear.
CAPTAIN: Okay eight, twelve, twenty.
FLIGHT ENGINEER: Lights out.
CAPTAIN: Set max power. Airspeed's alive.
COPILOT: Sixty knots.
CAPTAIN: My yoke [my control of takeoff].
COPILOT: Your yoke. Vee one. Rotate [leave ground].

Two seconds go by, after leaving ground at rotation.

CAPTAIN: Jesus Christ, help me on my yoke. Help me on the yoke. God. Help. Push forward.
COPILOT: I can't get it down.

At this point the aircraft is about 700 feet off the ground. Witnesses reported seeing an abnormally steep climb. Then the airplane abruptly banks left. It continues to roll to the left.

CAPTAIN: You got this damn thing in here.
ENGINEER: Come on, pull it. . . . Pull it back a little. Pull it back a little. Did you pull it back?
CAPTAIN: Okay, let me roll it into a bank.
COPILOT: What's the airspeed doing?
CAPTAIN: Damn it. Okay, come on. Get it over.
COPILOT: We're dead.
CAPTAIN: Lots of rudder, lots of rudder.
FLIGHT ENGINEER: Okay, it's clear now.
CABIN: [*Sound of ground-proximity-warning horn and recorded voice: "Whoop, whoop! Pull up, pull up!"*]
FLIGHT ENGINEER: Oh, god.

END OF TAPE

Flight 15 rolled and went straight into the ground between two hangars. The airplane exploded on impact, causing a severe fire.

What had caused the disaster seemed particularly tragic in light of the care and professionalism of the crew. A metal device called a gust lock—a T-shaped piece of aluminum four inches long and ten inches wide—was jammed in the copilot's control yoke. The retiring crew of Flight 15 had inserted the lock behind the control yoke on the copilot's side of the cockpit in order to raise the elevator-control surface on the horizontal stabilizer in the tail of the airplane during the cargo-loading operation. The gust lock was meant to be removed before takeoff, but the incoming copilot had failed to notice it. The captain, who was flying the takeoff, did not feel the jammed elevator until he had rotated the aircraft. By then those aboard were all but doomed. All three died. The hull-loss value was $8 million.

Carrollton, Georgia

August 21, 1995

■

Atlantic Southeast Airlines Flight 529

That day, just after noon, Atlantic Southeast Airlines Flight 529, a turboprop commuter flight, took off from Atlanta Hartsfield International Airport bound for Gulfport, Mississippi, with twenty-six passengers and a crew of three. The twin-engined Embraer EMB-120RT airplane, manufactured in Brazil, had flown for 17,151 hours in 18,171 takeoff-landing cycles.

At the time of departure, the clouds were at 200 feet scattered, with a ceiling of 1,600 feet broken, 3,400 feet overcast. Visibility was two miles. There was light rain and fog; temperature was 73 degrees F.; winds were light. The aircraft was carrying 350 gallons of fuel equally distributed in the left- and right-side wing tanks.

We pick up the CVR nine minutes after Flight 529 took off from Atlanta, climbing to 12,000 feet.

CAPTAIN: Tell Robin [the lone flight attendant in the back of the aircraft] it'll just be a couple of minutes [before the ride will] smooth out.

FLIGHT ATTENDANT: [*Coming to cockpit*] Hello.

COPILOT: Hey, Robin.

FLIGHT ATTENDANT: Hi.

COPILOT: It'll be just a couple more minutes like this; it'll smooth out.

FLIGHT ATTENDANT: Uh, a couple more minutes and then I can get up?

COPILOT: Yes, ma'am.

FLIGHT ATTENDANT: All right, thank you.
COPILOT: See ya.
ATLANTA CONTROL: Five twenty-nine, climb and maintain one three thousand [13,000 feet].
CAPTAIN: Thirteen. Props ninety.
COPILOT: [*On public-address system*] Ladies and gentlemen, good afternoon. Welcome aboard Atlantic Southeast Airlines Flight Five twenty-nine, service to Gulfport. We're passing through thirteen thousand feet. Captain has turned off the fasten-seat-belt sign. You are free to move about the cabin as you wish. . . . Gulfport on the hour is calling for partly cloudy skies, temperature of eighty degrees, and winds ten miles an hour out of the northeast. [If] there's anything we can do to make your flight more enjoyable, please do not hesitate to call upon us. And thank you for flying ASA.
CONTROL: Five twenty-nine, maintain one four thousand [14,000 feet]. Contact Atlanta Center on one three four point nine five.

In the next couple of minutes, Flight 529 climbs to 20,000 feet without incident, and the captain complains mildly about the repeated sounds of the altitude beeps. They have been flying now for nearly twenty-one minutes.

CABIN: [*Sound of several thuds*]

The torque on the left engine decreases to zero. The airplane rolls to the left, pitches down, and starts to descend at 5,500 feet per minute. In the cabin, the passengers hear a loud noise and the cabin shudders. Two of the three blades from the left propeller are wedged against the front of the wing. The flight attendant looks out the window and observes "a mangled piece of machinery where the propeller and the front part of the cowling was."

CAPTAIN: We got a left engine out. Left power lever. Flight idle.
CABIN: [*A shaking sound starts and lasts for thirty-three seconds*]
CAPTAIN: Left condition lever. Left condition lever.
COPILOT: Yeah.
CAPTAIN: Feather [propeller].
CABIN: [*Series of beeps for one second, indicating an engine fire*]

CAPTAIN: Yeah, we're feathered. Left condition lever. Fuel shutoff. I need some help here.
COPILOT: Okay.
CAPTAIN: I need some help on this.
COPILOT: [You said the propeller is] feathered?
CAPTAIN: Uh.
COPILOT: It did feather.
CAPTAIN: It's feathered.
COPILOT: Okay.
CAPTAIN: What the hell's going on with this thing?
COPILOT: I don't know. . . . [I] got this detector inop[erative].
CAPTAIN: Okay, let's put our headsets on. I can't hold this thing. Help me hold it.
COPILOT: Okay.
CAPTAIN: Okay, comin' on headset.
COPILOT: Atlanta Center, Five twenty-nine, declaring an emergency. We've had an engine failure. We're out of fourteen [thousand] two [hundred] feet at this time.
CONTROL: Five twenty-nine, roger. Left turn, direct Atlanta.
CAPTAIN: God damn.
CABIN: [*Sound of heavy breathing*]
CAPTAIN: All right. Turn your speaker off. Oh, we got it. It's . . . I pulled back, the power back.
CONTROL: Five twenty-nine, say altitude descending to.
COPILOT: We're out of eleven [thousand] six [hundred] at this time.
CAPTAIN: All right. It's . . . it's getting more controllable here. The engine . . . Let's watch the speed. All right, we['re] trimmed completely here.
COPILOT: I'll tell Robin what's going on.
COPILOT: [*On interphone to flight attendant*] Okay, we had an engine failure, Robin. We declared an emergency. We're diverting back to Atlanta. Go ahead and, uh, brief the passengers. This will be an emergency landing back in—
FLIGHT ATTENDANT: All right, thank you.
CAPTAIN: Tell 'em we want—
CONTROL: Five twenty-nine, can you level off or do you need to keep descending?
CAPTAIN: [Tell 'em] we ca— We're gonna need to keep . . . descending. We need an airport quick.

COPILOT: [*To Control*] Okay, we, uh, we're going to need to keep descending. We need an airport quick. Uh, roll the [emergency] trucks and everything for us.

CONTROL: Five twenty-nine, West Georgia, the regional airport, is at your . . . ten-o'clock position and about ten miles.

COPILOT: Understand—ten o'clock and ten miles.

CAPTAIN: Let's get the, uh, engine-failure checklist, please.

COPILOT: [*Getting the manual, paging through it*] Okay, I'll do it manually here. Okay, engine failure in flight. Power level's flight idle.

CONTROL: Five twenty-nine, you need to be on about a zero three zero heading for West Georgia Regional, sir.

COPILOT: Roger, we'll turn right. We're having, uh, difficulty controlling right now.

COPILOT: [*Reading manual again*] Okay, condition lever's feather.

CAPTAIN: All right.

COPILOT: [*Reading checklist*] Main auxiliary generators of the failed engine off.

CAPTAIN: Okay, I got that.

COPILOT: [*Reading*] APU [auxiliary power unit] . . . if available, start. Want me to start it?

CAPTAIN: We gotta bring this down. . . . Bring those. . . . Put that off.

CONTROL: Five twenty-nine, say your altitude now, sir.

COPILOT: Out of seven thousand . . .

CONTROL: Five twenty-nine, West Georgia Regional is your closest airport. The other one's, uh, Anniston. And that's about thirty miles to your west, sir.

CAPTAIN: [*To copilot*] How long, how far [is] West Georgia Reg[ional]? What kind of runway they got?

COPILOT: [*To Control*] What kind of runway's West Georgia Regional got?

CAPTAIN: [*To copilot*] Go ahead and finish the checklist.

CONTROL: West Georgia Regional is, uh . . . it's five thousand feet [long].

COPILOT: [*Going through the checklist*] Okay, APU started. Okay, prop sync, off. Prop sync's comin' off.

CAPTAIN: Okay.

COPILOT: Fuel pumps, failed engine. You want, uh, max on this?

CAPTAIN: Go ahead, please.

COPILOT: Okay.

CONTROL: . . . And it is asphalt, sir.

COPILOT: [*Continuing with checklist*] Hydraulic pump, failed engine? As required. Put it to the on position?
CAPTAIN: Correct.
COPILOT: Okay.
CAPTAIN: Okay.
COPILOT: Cross bleed open.
CAPTAIN: Okay.
COPILOT: Electrical load, below four hundred amps.
CAPTAIN: It is. . . . Well, you don't need to do that. Just leave that alone. All right, single-engine checklist, please.
CONTROL: Five twenty-nine, I've got you now and the airport's at your—say, say your heading now, sir.
COPILOT: Right now we're heading zero eight zero.
CONTROL: Roger, you need about ten degrees left. [The airport] should be twelve o'clock and about eight miles.
COPILOT: Ten left, twelve 'n' eight miles, and, uh, do we got a ILS to this runway?
CONTROL: Uh, I'll tell you what. Let me put you on Approach. He works that airport, and he will be able to give you more information. Contact Atlanta Approach on one two one point zero, sir.
CAPTAIN: We can get in [land] on a visual [approach without instruments].
CONTROL: Good luck, guys.
CAPTAIN: Engine's exploded. It's just hanging out there.
ATLANTA APPROACH: Five twenty-nine, Atlanta Approach.
COPILOT: Yes, sir. We're with you. Declaring an emergency.
CAPTAIN: We can get in on a visual. Just give us vectors [headings].
COPILOT: [*To Approach*] Just give us vectors. We'll go the visual.
CAPTAIN: [*To copilot*] Sing, single, single-engine checklist, please.
COPILOT: Where the hell is it [the checklist]?
APPROACH: Five twenty-nine, say altitude leaving.
COPILOT: We're out of nineteen hundred [1,900 feet] at this time.
CAPTAIN: We're below the clouds. Tell 'em. . . .
COPILOT: [*To Approach*] Okay, we're, uh, VFR at this time. Give us a vector to the airport.
APPROACH: Five twenty-nine, turn left, uh, fly heading zero four zero. . . . The airport's at your, about ten o'clock and six miles, sir. Radar contact lost [at] this time.
CABIN: [*Mechanical voice: "Five hundred (feet). Too low gear."*]

APPROACH: Five twenty-nine, if able, change to my frequency, one one eight point seven. The airport, uh, [is] in the vicinity of your ten o'clock at twelve o'clock and about four miles or so.
CAPTAIN: [*To copilot*] Help me. Help me hold it. Help me hold it. Help me hold it.
CABIN: [*Vibrating sound as the stick shaker starts warning of stall*]
COPILOT: Amy, I love you.
CABIN: [*Sound of grunting; sound of impact*]

END OF TAPE

While the flight attendant was preparing the passengers for an emergency landing, she looked out the window and saw treetops. She had already checked with all the passengers to see whether they understood how to brace themselves against impact, and as she took her own seat and belted herself in, she yelled instructions to the passengers until the time of impact.

The airplane sheared off several trees along a path 360 feet long before hitting the ground at a slight upslope. The airplane came to rest 850 feet from where it first hit the trees.

On impact, overhead storage bins fell down and the seats were torn from their anchors on the cabin floor. As the airplane slid on its left side, holes large enough to see through were punched in the fuselage.

About a minute after the airplane came to a stop, crackling sounds and sparks from electrical wires preceded a fire that quickly engulfed the cabin. Spilled aviation fuel saturated some of the passengers' clothing, and one passenger saw "a couple of people on fire."

None of the passengers escaped through the exit doors. Those who could manage escaped through the holes in the fuselage.

Despite second-degree burns to her ankles and legs, the flight attendant ran through the flames to help passengers out of the aircraft. She put out the flames on at least one person who was on fire.

The copilot tried to open the right cockpit window, which was damaged by the impact. He could not get out the cockpit door. He chopped a hole in the cockpit window with a small ax. He handed the ax out the hole to a passenger, who chopped at the window but could not make a hole large enough for the copilot to exit through.

A continuous roar of fire was heard in the cockpit from behind the cockpit door.

The copilot and captain were trapped.

Five minutes after the crash, the local fire department arrived. When firemen could not smash through the cockpit window with larger axes, they poured water in the hole to control the fire that was burning in the cockpit. Once they extinguished the flames in the cabin, they rescued the copilot through the cockpit door. The captain was already dead of burns and smoke inhalation. The copilot suffered burns over 80 percent of his body.

The captain and seven passengers died. The copilot, the flight attendant, and nineteen passengers survived the crash of Flight 529.

18

Nashville, Tennessee

February 3, 1988

■

American Airlines Flight 132

This incident presaged a disaster years later in the crash of the ValuJet in the Everglades (see Chapter 25). On February 3, 1988, a spontaneous chemical fire smoldered out of sight in the cargo hold of American Airlines Flight 132. On final approach to land, the cabin floor was softening, sinking, and simply melting under tremendous heat. A deadheading copilot riding in the passenger cabin who observed the floor told the copilot in the cockpit, "[We're] gonna have to land this thing in a hurry."

Until this moment, the flight was nothing but routine.

A McDonnell Douglas DC-9, Flight 132 had departed Dallas–Fort Worth at 2:45 that afternoon, destined for Nashville. There were 120 passengers, 4 flight attendants, and 2 deadheading flight-crew members aboard. The cargo holds contained 6,365 pounds of air freight. There was a 20-pound cylinder of oxygen and a 104-pound drum of textile-treatment chemicals, including 5 gallons of hydrogen peroxide solution, an oxidizer, and 25 pounds of a sodium-orthosilicate-based mixture, which is a granular material. No labels marked the drum to indicate the hazards of its contents. The oxygen cylinder and the drum were put in the middle cargo hold; the drum was laid on its side.

We pick up the CVR tape just as the flight attendant is telling the cockpit about the fire.

COPILOT: [*Speaking on interphone to back of the aircraft*] Hello.
FLIGHT ATTENDANT: [*Calling the cockpit on interphone from back of the aircraft*] Hi. We've got smoke in the cabin.
COPILOT: Okay.
FLIGHT ATTENDANT: We don't know where it's coming from. It's past the, ah, exit. [We've] got an H-two-O extinguisher. . . .
NASHVILLE APPROACH CONTROL: American One thirty-two, descend and maintain two thousand five hundred [feet].
CAPTAIN: Two thousand five hundred, American One thirty-two.
COPILOT: [*To captain*] We got smoke in the . . . ah . . .
FLIGHT ATTENDANT: It's a real bad smell.
COPILOT: [*To flight attendant*] Okay, I smell it up here now.
[*To captain*] [They] smell smoke. [They] got smoke in the cabin.
CAPTAIN: Smoke, or . . .
COPILOT: It smells electrical.
CAPTAIN: . . . fumes.
COPILOT: They can't tell where it's coming from. It smells electrical.
CAPTAIN: Is it fumes or smoke?
FLIGHT ATTENDANT: Yeah, a pilot [deadheading copilot who was riding in the passenger cabin] said the floor is getting really soft, and he said we need to land.
COPILOT: Okay. Who says the floor is getting soft?
FLIGHT ATTENDANT: [*Handing interphone to the deadheading copilot*] Here he is.
DEADHEAD COPILOT: Hey, boss.
COPILOT: Yes?
DEADHEAD: You got the floor back here in the middle . . . dropping out slightly.
COPILOT: Okay.
DEADHEAD: You['re] gonna have to land this thing in a hurry.
COPILOT: Okay, we're gettin' it down now.
DEADHEAD: Okay, be quick.
COPILOT: Okay.
DEADHEAD: Hey, have the [fire] trucks meet us [once we land].
COPILOT: [*To captain*] [We] have a flight officer back there, says that the floor is getting soft. [We] probably ought to drop the [landing] gear. There's somethin' going on in the, ah, floorboard.

CAPTAIN: Put the gear down.
CABIN: [*Sound of landing gear being lowered*]
COPILOT: [*To flight attendant*] Okay, now how far back is the floor gettin' soft?
FLIGHT ATTENDANT: Well, ah, the captain [deadheading copilot] is in the aisle right now. He's about midway through to—
COPILOT: About where the [landing] gear might be?
FLIGHT ATTENDANT: Yes.
COPILOT: Okay. Why don't you go back and buckle in.
FLIGHT ATTENDANT: We're all seated.
COPILOT: Okay, fine. [*To captain*] Okay, what do you want me to do here? Okay, seat belt [sign] . . .
CAPTAIN: Yes.
COPILOT: No smoking sign . . .
CAPTAIN: No smoke. Just fumes, right?
COPILOT: So far it's just smoke . . . fumes. [*To flight attendant on interphone*] You don't see any smoke. It's just fumes?
FLIGHT ATTENDANT: Bad fumes. Startin' to hurt my eyes.
COPILOT: Okay. I'm gonna get off the phone. Call me if anything important changes.
FLIGHT ATTENDANT: Okay.
CAPTAIN: [*To copilot*] Did you call the Tower?
NASHVILLE TOWER: American One thirty-two, Nashville Tower. Wind calm [on] Runway Two left. Cleared to land.
CAPTAIN: No problems.
COPILOT: There's just fumes back there.
CAPTAIN: We've had fumes before, from the APU is where [it came from] at least initially. Okay, we got [landing] gear.
COPILOT: Gear.
CAPTAIN: Spoiler lever, auto brakes. No. Flaps are good. Lights. Are we cleared to land?
COPILOT: [*To Tower*] American One thirty-two, are we cleared to land?
TOWER: Affirmative.
COPILOT: Roger. [*To captain*] Do you want to call any . . . [emergency equipment on the ground]?
CAPTAIN: We don't have any problems yet. Just a few fumes.
COPILOT: You don't smell it?
CAPTAIN: Yeah, I smell it.

COPILOT: You are cleared to land. Landing checklist is complete. Five hundred feet, sinkin' a thousand plus five. Four hundred [feet]. Three hundred. There's two hundred. One hundred. On the tape, fifty, forty, thirty, ten, five . . .

CABIN: [*Sound of touchdown*]

COPILOT: Reverse [thrust]. Hundred knots. Eighty knots.

TOWER: American One thirty-two, turn right. When able, contact Ground Control.

COPILOT: Sixty knots.

NASHVILLE GROUND CONTROL: American One thirty-two, Nashville Ground. Roger, your option [is] to enter Tango Two [runway exit] or come down to Tango Four. Advise.

COPILOT: [*To captain*] Tango Two or Tango Four.

CAPTAIN: Ah, let's see. . . .

COPILOT: This is my first time in here, so let me look this up.

CABIN: [*Sound of cabin chimes with flight attendant calling cockpit*]

COPILOT: I'm here.

DEADHEAD COPILOT: [*On interphone*] You've got a big problem back here, and I'm not sure if you . . . The problem is, I don't know where the heat is comin' from. It's comin' up through the floor.

COPILOT: Do you see any smoke?

DEADHEAD: Yeah, there's smoke. Just a little bit.

COPILOT: Okay, okay.

DEADHEAD: We better get outta here.

COPILOT: Okay.

FLIGHT ATTENDANT: [*To captain*] Ah, Captain?

COPILOT: [*To captain*] There's a crew[man] back there that says we better get outta here. He says there's smoke comin' through the floor.

FLIGHT ATTENDANT: I don't see it [the smoke]. We had a first officer here with us. He's the one. He's been checkin' the floor. He's in uniform. That's who you've been talkin' to. . . .

COPILOT: [*To captain*] She don't see [the smoke].

FLIGHT ATTENDANT: He [the deadhead copilot] thinks it's real soft, the floor's real soft.

COPILOT: [*To captain*] The floor is getting very, very soft.

CAPTAIN: Okay, let's get out of here. Call Ground. . . .

COPILOT: [*To flight attendant*] Okay, we're gonna get out of the airplane now.

FLIGHT ATTENDANT: Okay, Easy Victor [evacuation].
COPILOT: Ah, stand by.
FLIGHT ATTENDANT: Okay.
CAPTAIN: Give me the checklist.
COPILOT: [*To Ground Control*] Ah, roger, sir, would you call out the fire equipment? We've got the possibility of some fire, some real hot stuff, in the cargo compartment. The floor is real hot. We're gonna get 'em [the passengers] out.
GROUND CONTROL: Okay, we got 'em on the phone, American One thirty-two.
COPILOT: [*To captain*] Okay, ground evac. Ah, Tower. Called the Tower. Flaps.
CAPTAIN: [Flaps] forty [fully extended].
CAPTAIN: Spoiler lever . . .
CAPTAIN: You get out of here. You go help [the flight attendants]. Retract brakes. Park fuel levers.
COPILOT: Cutoff . . .

END OF TAPE

The captain ordered the evacuation two minutes and six seconds after Flight 132 touched down, and the inflatable slides were deployed at the two forward cabin doors, the aft galley door, and in the tail cone. The overwing exits were not used. No instructions were given to the passengers over the public-address system. Neither were they prepared for the evacuation before landing. During the evacuation, the flight attendants shouted commands at the passengers: "Unfasten seat belts" and "Come this way" and "Remove shoes" and "Don't take anything with you."

After the passengers had safely evacuated the airplane, an American Airlines maintenance employee on the ground asked the captain about the problem. The captain said there was a fire in the cargo area. They opened the aft cargo compartment and saw a little smoke inside. Then they opened the middle cargo compartment. Thick, white-gray smoke poured out.

The Tower's call dispatched fourteen fire fighters with six vehicles, four crash-fire rescue units, and two quick-response units to the aircraft, which had pulled to a stop on the apron beside the runway. The emergency units sprayed about 120 gallons of water into the middle cargo compartment to douse the smoldering fires. Neither aqueous film-forming foam nor dry chemicals to fight fires was used.

None of the 126 crew and passengers was injured seriously; 9 passengers and 4 crew suffered minor injuries.

19

Pine Bluff, Arkansas

April 29, 1993

■

Continental Express Flight (JetLink) 2733

This was a flight no one would have chosen to take, not because any of the twenty-seven passengers and three crew members was killed or even injured, but because the twin-engined propeller Brazilian Embraer EMB-120 RT stalled suddenly while climbing to its flight altitude at 17,000 feet and plummeted to 5,500 feet before the crew regained control. Then, with one of the two engines badly damaged and with three of the four propeller blades missing, the airplane could not maintain level flight. In other words, it was going down and only an airport nearby could save it. What made this a particular flight from hell, the crew could have avoided the whole incident if the pilot had not been—in the characterization of the National Transportation Safety Board—so "relaxed."

Continental Airlines' Continental Express Flight 2733 (called Jet-Link 2733) left Little Rock, Arkansas, en route to Houston, Texas, in the middle of the afternoon. The flight was instructed to climb through 7,500 feet and maintain a flight level of 22,000 feet. The weather was overcast, with a visibility of nearly five miles, temperature of 68 degrees F., light winds, moderate icing in the clouds, and precipitation between 12,000 and 20,000 feet. In other words, the weather was good for flying.

After takeoff and while climbing, the copilot wrote in his logbook and ate his in-flight meal. The captain, who had a reputation for being "easy to get along with," loosened his safety harness and put his foot up on the console, according to one passenger who was watching through the open door. The flight attendant had made her way forward and was standing in the door chatting with the captain, who had put the aircraft on autopilot. The crew had spent the morning and early part of the afternoon together poolside at the motel where they were staying in Little Rock.

We will begin the tape transcript during takoff.

LITTLE ROCK TOWER: JetLink Two seven three three, [you are] cleared for takeoff [on Runway] Two two right. Turn left [after takeoff on] heading one eight zero.

COPILOT: Turn left [to] one eight zero, cleared to go [on] Two two right, Two seven three three.

CABIN: [*Sound of increasing engine noise as throttles are pushed forward*]

COPILOT: [Throttles] set. Eight [mph]. Vee one. Vee two. Positive rate.

CAPTAIN: [Landing] gear up.

COPILOT: Acceleration.

CAPTAIN: Okay, flaps [up].

TOWER: JetLink Two seven three three, contact Departure.

COPILOT: Two seven three three, so long. Departure, JetLink Two seven three three's with you [on radio frequency] one point three, [flying at] four thousand [feet], [at] one eight zero heading.

LITTLE ROCK DEPARTURE: JetLink Two seven three three . . . maintain one zero thousand [10,000].

CAPTAIN: What do you think you're doin'?

COPILOT: I don't know.

For the next few minutes Flight 2733 continues to climb through clouds, with the copilot contacting Little Rock Departure and Memphis Center for heading and altitude instructions. The copilot tells the captain that he is going to run some calculations to see whether a flight altitude of 26,000 feet might not be better for them than the assigned 22,000 feet.

COPILOT: I'm gonna go ahead and run a weight for two six zero right.

CAPTAIN: I don't . . . I don't care. I don't know what the winds are like up there.

The flight attendant, while the aircraft is still climbing, comes to the cockpit door.

FLIGHT ATTENDANT: Hi.
CAPTAIN: [*To copilot*] Not much difference [in the winds], huh?
COPILOT: At eighteen [thousand feet], it's, ah, two ninety at twenty-six, at two four [24,000 feet], it's twenty-six at thirty-three.
CAPTAIN: It [the flight altitude's being between 22,000 and 26,000] really don't make a big difference.
COPILOT: The higher you go . . .
CAPTAIN: . . . the more it turns into a headwind anyway.
COPILOT: Yeah. What do you want to go for a final [cruising altitude]? You want to go [to 26,000 feet]?
CAPTAIN: I don't care.

The flight attendant asks if the captain and the copilot want to eat. The copilot says he does, the captain says, "No, no, thanks." And the flight attendant leaves the cockpit momentarily to get the meal for the copilot.

CAPTAIN: [*To copilot and flight attendant*] Do you guys really want to eat somethin' that bad?
COPILOT: [*To captain*] That girl has got you all screwed up. . . . It looks like [you can't] concentrate.

The captain apparently has been put on a diet by "that girl," and he is worried about maintaining his diet.

CAPTAIN: Man, I just couldn't help it. I was tryin'. I was tryin' to change the conversation, all right.
COPILOT: Yeah.
MEMPHIS CENTER: JetLink Two seven three three, climb and maintain flight level two three thousand [feet].
CAPTAIN: . . . When I get back to [one hundred] eighty-five [pounds].
COPILOT: Eighty-five . . .

The copilot and the captain engage in a bit of banter, and when the flight attendant comes back into the cockpit, she and the captain take up the conversation, now about a bug that is stuck on the windshield of the aircraft.

FLIGHT ATTENDANT: [*To captain*] Oh, do the windshield wipers. That'll wipe it [the bug] off.
CAPTAIN: Naw. I can't. Naw. I can't do it 'cause we're goin' too fast.

FLIGHT ATTENDANT: That'd be funny. That'd be funny. That'd crack me up. That'll make my day.
CAPTAIN: . . . Off.
FLIGHT ATTENDANT: Do it, do it, do it.
CAPTAIN: Okay.
FLIGHT ATTENDANT: Can't we climb any faster?
CAPTAIN: Why do you want to get up so fast?
FLIGHT ATTENDANT: . . . I can't pull it uphill.

The flight attendant wants the captain to reach cruise altitude faster because she does not want to push the drinks cart up the aisle while the airplane is still in a climb attitude.

CAPTAIN: Okay.
FLIGHT ATTENDANT: The last time I had to pull it up the hill [aisle].
CAPTAIN: We'll try to get up a little more [a little faster].
FLIGHT ATTENDANT: We'll try to get up there?
CAPTAIN: Yeah. We're almost there [to cruise altitude]. Another six thousand feet, another six minutes.

Here the flight attendant fools with the captain. She offers him something that the tape does not make clear.

FLIGHT ATTENDANT: Gimme, gimme, gimme. Climb. Gimme, gimme, gimme.
CAPTAIN: [*To flight attendant*] You're right. Better to receive than to give.
FLIGHT ATTENDANT: What?
CAPTAIN: Ah, like you said. Better to receive than to give.
FLIGHT ATTENDANT: I don't know.
CAPTAIN: [I] live by that.

The captain and the flight attendant engage in another five seconds of what the NTSB calls "nonpertinent" conversation.

CAPTAIN: We're not climbin' very fast. Yeah, we're really heavy.

At this point, ice is starting to form on the wings, and the improper instructions that the captain gave to the autopilot for flight vertical mode are making the airplane unstable. Its "heaviness" is the indication that it is about to stall.

FLIGHT ATTENDANT: I know . . . I know [I couldn't, or you couldn't] walk, like walk a straight line [and] not sway all over it.
CAPTAIN: [*Laughs*]
FLIGHT ATTENDANT: [*Laughs*]
FLIGHT ATTENDANT: So that's what you meant by a weight problem.
CAPTAIN: [*To copilot*] Frank, hang on. This ain't right.

Two seconds go by. The stall-indicating stick shaker starts to shake the control yoke, and the autopilot automatically disconnects. The airplane plummets. The flight attendant is thrown back into the passenger cabin. During the ensuing descent—free fall of the airplane—the flight attendant yells to the passengers to assume "impact" position; she tells them the location of the emergency exits. She tells them that their landing will be "hard and fast."

CAPTAIN: Airspeed.
CABIN: [*Stall warning sounds*]
CAPTAIN: Hang on. Hang on. Power's up . . . power's . . .
CABIN: [*Sound of increasing engine noise*]
CAPTAIN: Up . . .

Here the captain pulls back hard on the control column, making the recovery of the airplane more difficult to achieve as it falls nearly 12,000 feet.

CABIN: [*Sound of vibration*]
CAPTAIN: Up . . . I think it's an overspeed. It's an overspeed.

END OF TAPE

As the tape ended, the captain shut down the left engine, which he thought was racing out of control, when in fact it was badly damaged. The copilot lowered the landing gear, which helped them to regain control of the airplane, and the crew declared an emergency, stating that they needed to land "immediately and were losing altitude." The captain decided to make the emergency landing at Pine Bluff Airport. Five minutes before Flight 2733 reached the airport, the Pine Bluff airport manager ordered Runway 17 cleared of construction equipment. Flight 2733 broke out of the clouds about one mile from the end of the runway. The captain overshot the runway and touched down with 1,880 feet of runway remaining. He applied the brakes, but the runway was wet, and the airplane "hydroplaned," or skidded, off the asphalt surface onto the grass, narrowly missing a construction vehicle and several construction workers who were watching at the time.

The airplane came to rest 687 feet from the end of Runway 17. The crew and passengers immediately evacuated. There was only one minor injury.

The NTSB determined that the probable causes of the incident were "the captain's failure to maintain professional cockpit discipline, his consequent inattention to flight instruments and ice accretion, and his selection of an improper autoflight vertical mode, all of which led to an aerodynamic stall, loss of control, and a forced landing."

Anchorage, Alaska

March 31, 1993

■

Japan Airlines Flight 46E

As they taxied out for takeoff from Alaska International Airport, en route to Chicago-O'Hare, the crew of Japan Airlines Flight 46E (a cargo flight) were warned of severe turbulence, and later, just before takeoff, they were told that a flight departing ahead of them had experienced severe turbulence at 2,500 feet while climbing out from Runway 6R. Notice the differences and similarities between this flight and El Al 1862 (Chapter 10): the same kind of airplane with the same catastrophic failure.

Flight 46E was a Boeing 747-121 loaded with cargo, with a crew that consisted of a captain, a copilot, a flight engineer, and two company employees along for the ride. Other aircraft in the area or at the airport were Japan Airlines 42E and Peninsula 4205.

We pick up the CVR just as the flight is taxiing out for takeoff.

FLIGHT ENGINEER: You gotta get your priorities right.
CAPTAIN: Yeah, right. What'd you come up here for?
FLIGHT ENGINEER: This is [the best angle or view] right here that I can see.

The flight engineer apparently is taking a photograph through the cockpit window of two airplanes taking off ahead of them.

COPILOT: [Try to] get both of them in the same picture.
FLIGHT ENGINEER: I am.
CABIN: [*Laughter*]

CAPTAIN: Do it.
FLIGHT ENGINEER: Oh, yeah, I got it, actually.
CABIN: [*Laughter*]
FLIGHT ENGINEER: Oh, I got it twice. That's cool. I didn't even see what you're talking about. I won't get the rotate, but those are good pictures. Actually I got more of the little guy than I had of our plane, but it's kinda neat for comparison.
CAPTAIN: Fun stuff, huh?
FLIGHT ENGINEER: Really. Okay, uh, let's see, where were we? We did, uh, stab[ilizer] trim, then we did INS [Inertial Navigation System]. How about pitot heat?
COPILOT: On.
FLIGHT ENGINEER: Fuel heat's checked and off. Anti-ice?
COPILOT: Checked and off.
FLIGHT ENGINEER: Shoulder harness?
CAPTAIN: On.
COPILOT: On.
FLIGHT ENGINEER: Comin' on. Takeoff brief?
COPILOT: Understood.
CAPTAIN: Same brief as before. Ah, the departure is still eleven miles, two thousand feet, three-hundred-and-thirty-degree heading, going two zero zero. Okay?
FLIGHT ENGINEER: Yup. Understood. Flight instruments?
CAPTAIN: Checked.
COPILOT: Checked.
FLIGHT ENGINEER: Taxi check complete. That was a cool picture, though. I didn't even see the other guy comin', and I was just looking because all I saw was the little tube, you know, and, uh, and what I saw [in] my viewfinder. All of a sudden he came in there and . . . there's a [C–]One forty-one piece of crap or a C-Five, one of the two.
COPILOT: Are you still flying those?
FLIGHT ENGINEER: Yeah, I hate 'em.
CAPTAIN: [It's called a] Skylifter.
FLIGHT ENGINEER: It's gross . . . the whole air-conditioning system stinks. [When you] get off the airplane you smell like a [C–]One forty-one. Yeah, that's what it is.
CAPTAIN: [Somebody] had a model of a One forty-one in there and on the box it said, "Sky Lizard."
FLIGHT ENGINEER: Did it really? Okay, flaps, Vee speeds, trim?

CAPTAIN: Rechecked for Six right.

COPILOT: Rechecked for Six right.

FLIGHT ENGINEER: Checked. [Six right] for takeoff. Six right anti-ice? Off, INS?

CAPTAIN: Checked.

FLIGHT ENGINEER: APU goin' off. Fuel system set for takeoff. Forty-two Echo said expect a rough ride.

COPILOT: Japan Air Forty-six Echo heavy is ready.

ANCHORAGE TOWER: Japan Air Forty-six Echo heavy, Anchorage Tower. Runway Six right, taxi into position and hold. Pilot reports severe turbulence leaving two thousand five hundred. . . .

CAPTAIN: Okay.

COPILOT: Roger, understood. Thank you.

FLIGHT ENGINEER: Batten down the hatches, folks. We're expecting a rough ride.

TOWER: Japan Air Forty-six Echo heavy, Runway Six right, cleared for takeoff.

FLIGHT ENGINEER: [Reports of] severe turbulence on climb out. I don't [know] what else is out on that galley now, but we're getting ready to blast off, so just keep an eye out.

CAPTAIN: All the way on the checklist.

FLIGHT ENGINEER: All the way.

CAPTAIN: Cleared to go.

The crew completes its before-takeoff checklist.

CABIN: [*Sound of engine power increasing*]

CAPTAIN: Max power.

FLIGHT ENGINEER: Max power. Max power is set, and you've got . . . ninety-three percent.

CAPTAIN: Thanks.

COPILOT: Eighty knots. Vee one. Rotate. Vee two. Positive rate.

CAPTAIN: Gear up.

TOWER: Japan Air Forty-six Echo heavy, contact Departure.

COPILOT: Good day.

ANCHORAGE DEPARTURE: Japan Air Forty-six Echo heavy, turn right heading three five zero.

COPILOT: Good afternoon, Japan Air Forty-six Echo, out of one thousand for two zero zero.

DEPARTURE: Japan Air Forty-six Echo heavy, Anchorage Departure, radar contact, expect severe turbulence [at] two thousand five hundred. . . .
COPILOT: Roger.
CAPTAIN: Max climb power.
FLIGHT ENGINEER: Max climb.

Now the Boeing 747 enters the area of severe turbulence.

FLIGHT ENGINEER: Hang on, guys.
COPILOT: Left three three zero.
CAPTAIN: Flaps five.
COPILOT: Flaps five.
CAPTAIN: Flaps up.
FLIGHT ENGINEER: Flaps up. Whoa, whoa, whoa.
CABIN: [*Sound of snap and sound of warning horn*]

The violent turbulence throws the Boeing 747 into a severe banking turn of approximately 50 degrees. While the desired airspeed is 183 knots, the airspeed fluctuates about 75 knots from a high of 245 knots to a low of 170 knots. The number-two throttle slams to its aft stop, and the number-two thrust reverser shows deployment. The airplane pitches and rolls severely before the engine separates from the wing.

FLIGHT ENGINEER: Whoa, whoa, thrust reverser . . . We got auto-fail. We lost something.
COPILOT: [We] lost number one and two [engines].
CAPTAIN: [Number] two's gone.
FLIGHT ENGINEER: Number-two engine shut down. Whoa, whoa.
CABIN: [*"Bank angle, bank angle."*]
CAPTAIN: All right. [Declare an] emergency.
FLIGHT ENGINEER: Number-two start levers cut off.
DEPARTURE: Japan Air Four six Echo heavy, ah, Elmendorf Tower said that something large just fell off your airplane.
CAPTAIN: Yeah, we know it.
COPILOT: We know that. Ah, we're, ah, declaring an emergency.
DEPARTURE: Japan Air Four six Echo heavy, will you need to return to Anchorage?
CAPTAIN: We are returning to . . .

Shortly after the engine separates from the airplane, the flight crew declares an emergency, and the captain initiates a large-radius turn to the left to return and land on Anchorage International's Runway 6R. Maximum power is applied to the number-one engine. As the Boeing 747 maneuvers for landing approach, its bank angles exceed 40 degrees.

COPILOT: Stand by. Returning. And we are declaring an emergency.
DEPARTURE: Japan Air Four six Echo heavy, turn left heading two four zero, maintain three thousand, vector ILS Runway Six right final approach course.
CABIN: [*"Bank angle, bank angle."*]
FLIGHT ENGINEER: Okay, hang on. Do you wanna dump fuel?
COPILOT: Stand by one, stand by one. We've got an airplane comin' to us.
FLIGHT ENGINEER: Verify number-two fire handle.
CAPTAIN: Yes, sir, there's traffic ten o'clock, two miles, three thousand eight hundred, climbing rapidly.
CABIN: [*Sound of warning horn*]
CAPTAIN: We got traffic over . . .
FLIGHT ENGINEER: Number two, verify, fire handle.
DEPARTURE: Japan Air Four six Echo heavy, that traffic's leaving five thousand five hundred.
COPILOT: Number two [fire handle] set.
FLIGHT ENGINEER: We're turning back [to] Anchorage. We got two leading-edge devices out on the left side.
DEPARTURE: Japan Air Four six Echo heavy, descend and maintain one thousand six hundred. Can you use Runway One four? It's closer.
CAPTAIN: You dumping fuel?
COPILOT: [*To Departure*] Roger, stand by one, please.
FLIGHT ENGINEER: [*To captain*] Not yet . . . Need some help up here.
CAPTAIN: Okay, now. Okay, give me [the] flaps back.
COPILOT: Five, five?
CAPTAIN: Five.
COPILOT: Flaps coming five.
FLIGHT ENGINEER: Okay, fuel's coming off. Okay, flaps are coming down. Number-two engine's secured. Fuel dumping is in progress. You wanna dump down to five eighty-five, correct?
CAPTAIN: We're gonna, well, I'm having a real hard . . .

COPILOT: I think we lost the engine.
FLIGHT ENGINEER: We lost number-two engine.
DEPARTURE: Japan Air Four six Echo heavy, say your intentions.
FLIGHT ENGINEER: [We need] thirty minutes [to] dump [fuel].
COPILOT: We, we are, ah, we are going to maintain this heading to, ah . . . we are having problems with our flight controls, and also, ah, speed. Stand by one.
CAPTAIN: Everybody secured?
FLIGHT ENGINEER: Everybody's secured.
DEPARTURE: Roger, any runway, Japan Air Four six Echo heavy. Ah, any runway you need.
FLIGHT ENGINEER: Tell me what kinda' gross, what kinda' fuel weight that is. [I gotta find] zero fuel [weight].
CAPTAIN: Are you dumping fuel?
FLIGHT ENGINEER: Yes, we are. Would you like the quick return?
CAPTAIN: Quick return to the line, please.
FLIGHT ENGINEER: Okay, quick return, to the line. Landing gear?
COPILOT: Off, uh . . .
CAPTAIN: Off.
FLIGHT ENGINEER: You got all but number two [engine], number three is down, it's in progress. Okay, quick return. Landing gear?
COPILOT: Ah, up, a . . .
FLIGHT ENGINEER: Flaps?
COPILOT: Flaps are maintaining five degrees.
FLIGHT ENGINEER: Five degrees . . . speed-brake handle?
COPILOT: [Forward].
CAPTAIN: Did you declare an emergency for us?
DEPARTURE: Japan Air Four six Echo heavy, are you able to maintain terrain clearance?
COPILOT: Affirmative at this time, and, ah, we are maintaining, ah, thirteen hundred [feet].
DEPARTURE: Japan Air Four six Echo heavy, roger.
COPILOT: We are dumping fuel.
DEPARTURE: Japan Air Four six Echo heavy, roger.
FLIGHT ENGINEER: Okay.
CAPTAIN: You did declare an emergency?
COPILOT: Yes, affirmative.
CAPTAIN: Give me some speeds quickly.
FLIGHT ENGINEER: . . . Okay, speed's gonna be . . . for five eighty-five, I believe one forty-five [mph].

CAPTAIN: All right. How's our weight doing?

FLIGHT ENGINEER: I'll get it to you in a second. Don't have time to do it perfect.

CAPTAIN: Give me, ah, ILS on the, ah, radios.

FLIGHT ENGINEER: What was the fuel weight that you came up with? We need it now.

CAPTAIN: We're gonna lose number, number one [engine] here for temperatures.

CABIN: [*Sound of warning horn*]

DEPARTURE: Japan Air Four six Echo heavy, I have two F-Fifteens off your fight wing three miles, they have you visually, and if you'd like any panel inspection they said they [can] get in closer and, ah, and look you over.

CAPTAIN: Yeah, [tell them to] go ahead.

COPILOT: Go ahead.

DEPARTURE: [*To fighter jets*] Lion One, they said, ah, you can proceed on in and take a look and see how much damage has been done.

FLIGHT ENGINEER: Okay.

COPILOT: [*To captain*] Okay, which runway do you want?

CAPTAIN: [We're gonna] want [Runway] Six right.

COPILOT: [*To Departure*] We'd like the Runway Six right. This is, ah, Japan Air Forty-six Echo.

FLIGHT ENGINEER: In-flight failure check, shutdown check is complete.

CAPTAIN: Why am I losing airspeed here?

DEPARTURE: Japan Air Four six Echo heavy, understand turning inbound to Runway Six right.

FLIGHT ENGINEER: What's the airspeed?

CABIN: [*Recorded altitude warning: "One thousand feet."*]

COPILOT: You want, ah, some, ah, some rudder trim? Rudder trim?

CAPTAIN: That's not gonna help. Let's try to get this thing turned around.

FLIGHT ENGINEER: These numbers are wrong. One forty-five, one fifty-one. I don't know, it's back here [someplace]. [I had a] piece of paper. . . . No, I mean on, like, departure, I had that fuel-oil schedule.

CAPTAIN: I just want to stay out of the water here, you guys.

COPILOT: Okay.

CAPTAIN: Emergency power.

COPILOT: Okay, I got it.

CAPTAIN: . . . Flaps. Well, no, we can't.

COPILOT: You want flaps?
CAPTAIN: No.
FLIGHT ENGINEER: Hundred feet . . . fuel weight of seven hundred, right now.
CAPTAIN: How much [does] the airplane weigh?
FLIGHT ENGINEER: The airplane right now weighs seven hundred, your Vee speeds for flaps thirty—ref speed is . . . one sixty-three—Vee ref.
CAPTAIN: How much does the airplane weigh?
FLIGHT ENGINEER: I'm showin' it weighs six eighty-seven, right now. [There's] fuel dumping in progress.
CAPTAIN: [We will have] to land heavy.
FLIGHT ENGINEER: Okay, land heavy, heavy-weight landing for . . .
DEPARTURE: Understand, ah, he's lost the, ah, left [flap] on the left wing?
COPILOT: We got a fighter [jet] over out, over here. He's looking at, he's looking at us, yeah.
CAPTAIN: All right, we're going to get [no] left flap. All right, quick return, below the line.
DEPARTURE: Japan Air Four six Echo heavy, you've lost, ah, approximately fifty percent of the leading-edge slats on the left wing, and structural damage to the trailing-edge flaps.
CAPTAIN: All right.
COPILOT: You want to land [on this runway]?
CAPTAIN: I want to land on that runway, right now.
COPILOT: You want flaps, more flaps?
DEPARTURE: Japan Air Four six Echo heavy, Runway Six right, cleared to land.
CAPTAIN: All right. All right. We're landing. . . .
COPILOT: We're on [Runway] Six left.
CAPTAIN: Discontinue dump[ing of fuel].
FLIGHT ENGINEER: All right, discontinuing dump. And you want gear down?
COPILOT: Okay, slow down. You can slow down now.
CAPTAIN: Gear down.
COPILOT: Gear down?
CABIN: [*"Glide slope."*]
DEPARTURE: Japan Air Four six Echo heavy, did you copy the windshear report, sir?
FLIGHT ENGINEER: Landing gear and tilt?

COPILOT: Okay, do you want more flap, or flap five is okay?
CAPTAIN: We're gonna go . . .
FLIGHT ENGINEER: Landing gear tilt, down and green.
COPILOT: [*To Departure*] Roger, we copy. We are coming for Runway Six right.
CABIN: [*"Glide slope, glide slope."*]
DEPARTURE: Japan Air Forty-six Echo heavy, all the gear appears to be good.
COPILOT: [*To Departure*] Thank you.
FLIGHT ENGINEER: You got ADF on both, flight instruments, radio no flags, air condit—
CABIN: [*"Terrain, terrain."*]
CAPTAIN: Zero trim.
COPILOT: Two hundred feet. Zero the trim.
FLIGHT ENGINEER: Speed-brake handle.
CAPTAIN: Flaps twenty-five.
COPILOT: Flaps are comin' twenty-five. Flaps is twenty-five.
CABIN: [*"One hundred (feet) . . . fifty (feet) . . . thirty (feet)."*]
FLIGHT ENGINEER: Before-landing checklist complete.
CABIN: [*"Twenty (feet) . . . ten (feet)."*]
FLIGHT ENGINEER: Hang on, guys. Spoilers extended, reverse available one two three [engines], or one three four [engines].
DEPARTURE: [*To fighter jets*] Lion One, advise when he's down. Lion One, wilco and thank you for the assistance.
CAPTAIN: Thank you.
FLIGHT ENGINEER: Thank you. I can't see your speed so, uh . . .
COPILOT: Ninety [mph].
FLIGHT ENGINEER: Eighty knots.
CAPTAIN: Tell that guy thanks for his help.
FLIGHT ENGINEER: That's cool. Okay, we've got thrust reverser, lights out.
CAPTAIN: All right, secure . . .
COPILOT: Ah, thank you very much, Tower, this is Evergreen, ah, Japan Air Forty-six Echo. Thank the fighters for us.
DEPARTURE: Japan Air Four six Echo heavy, and, ah, they [the fighters] wish to say you did a good job and, ah, thank you.
FLIGHT ENGINEER: [*Laughs*]
DEPARTURE: Japan Air Four six Echo heavy, contact Anchorage Ground one two one point niner.
COPILOT: Thank you and roger, good day.

FLIGHT ENGINEER: [*To copilot*] Thanks, buddy. I don't care how much, you, I, how many beers I owe you in the past. This one I'm going to pay off on. Okay?

COPILOT: Okay, we just cleared the runway.

ANCHORAGE GROUND CONTROL: Japan Air Four six Echo heavy, taxi to parking. Say your gate number.

COPILOT: Roger, Romeo ten, and, ah, it's very, very, extremely heavy turbulence, on, ah, our takeoff on [and] the left turn.

GROUND CONTROL: Roger, taxi to Romeo ten.

CAPTAIN: We ripped off some flaps and stuff.

FLIGHT ENGINEER: We did some damage. . . .

CAPTAIN: [*Looking out window at wing*] Lots of parts missing out there.

FLIGHT ENGINEER: Good job, guys, both of ya.

CAPTAIN: Thank you for your help. Sorry we got disorganized.

FLIGHT ENGINEER: Okay, should we do some of the others, after landing? I'm gonna turn the probe heat and all that kind of stuff off, though.

CAPTAIN: Yeah, go ahead. I got the spoilers. . . .

FLIGHT ENGINEER: That was cool, that fighter out there takin' our pictures.

CAPTAIN: Did we lose number-two engine?

FLIGHT ENGINEER: Yes, we did.

COPILOT: No . . . huh? Yeah, wait a minute.

FLIGHT ENGINEER: We lost it. We lost number-two engine.

CAPTAIN: I had a hard time gettin' this thing trimmed enough. . . .

FLIGHT ENGINEER: Jeez, that bank angle and stuff, man, that was like crazy.

CAPTAIN: I was goin' full full right rudder to recover the bank.

FLIGHT ENGINEER: I admit it now. I was scared.

CAPTAIN: We were all scared.

More than three minutes go by as they taxi to a stop at the terminal.

FLIGHT ENGINEER: Okay, somebody can kiss me and tell me I'm still here. . . .

CAPTAIN: I'll kiss you in a minute here. . . .

END OF TAPE

The number-two engine pylon had separated from the wing, taking the engine with it and damaging the wing slats and flaps. Load forces created by severe or possibly extreme turbulence caused the failure of the pylon, which was later tested and discovered to have contained a weakening metal-fatigue crack.

No one on board the airplane or on the ground was injured.

Roselawn, Indiana

October 31, 1994

■

American Eagle Flight 4184

This was the fatal accident in the farmland of northern Indiana that made aviation authorities and the flying public seriously question commuter airline safety; it was the third crash involving a popular commuter airplane made by a French-Italian consortium.

All airplane accidents that result in fatalities are atrocious, of course, but this one seemed more so, if for no other reason than the way the victims were described by the coroner who examined them when he wrote that the passengers, all of whom died, had suffered "multiple anatomical separations secondary to velocity impact of aircraft accident." DNA had to be used to identify the bodies of the captain and copilot. Local authorities declared the crash site, which spread over twenty acres of a sodden soybean field, a "biohazard area," with all that that term implies.

American Eagle Flight 4184, a twin-engined propeller-driven Avions de Transport Regional 72-212 (ATR 72), boarded late at Indianapolis Airport under threatening skies. Most of the flight's sixty-four passengers were returning home to Chicago after a day of doing business in the Indianapolis area, or they were planning to catch connecting flights out of O'Hare. One passenger was worried about missing his connecting international flight to Frankfurt, Germany. Flight 4184 was due to leave Indianapolis at 2:10 P.M. (to arrive at Chicago-O'Hare at 3:15), but because of worsening weather at Chicago-O'Hare, the flight left the Indianapolis gate at 2:14 and was held on the ground for forty-two minutes before taking off. At 2:53, the ground controller at Indianapolis advised the crew of Flight 4184 "to expect a little bit of holding in the air."

At the time of the flight there was no particular warning about the icy conditions below 19,000 feet. The NTSB concluded that "there were no additional reports of any significant weather phenomenon in the vicinity" of Flight 4184's holding pattern. But, as the flight circled in the holding pattern, ice built up little by little on the aileron assembly out of reach of the deicing mechanisms. This ice would eventually reach a thickness that caused the airplane to roll over and plummet to the earth.

We pick up the CVR as Flight 4184 begins its turn into the holding pattern for Chicago-O'Hare.

CABIN: [*Sound of music*]

FLIGHT ATTENDANT: [I thought] I'd come to see what's going on up here.

CABIN: [*Sound of loud music*]

FLIGHT ATTENDANT: Is that like stereo radio? You don't have a hard job at all. We're back there [in the passenger cabin] slugging it out with these people [passengers].

CAPTAIN: Yeah, you are. We do have it pretty easy. I was telling Jeff [the copilot], I don't think I'd ever want to do anything else but this.

COPILOT: I . . . I like dealing [with] people in a way. It's kinda neat to be able to talk to them.

FLIGHT ATTENDANT: How late are we going to be?

CAPTAIN: Well . . .

FLIGHT ATTENDANT: We already got two people [passengers] that have already missed their [connecting] flights.

CAPTAIN: Oh, really?

FLIGHT ATTENDANT: Three-fifteen [P.M.] is one of them. [Flight 4184 was originally scheduled to land at 3:10 P.M.].

CAPTAIN: Three-fifteen, three-fifteen?

FLIGHT ATTENDANT: [*Joking*] [And] it's all your fault. Uh-huh. We weren't due in Chicago until three-fifteen.

CAPTAIN: Three-fifteen Eastern Time.

Passengers knew they might miss their connecting flights at Chicago-O'Hare before they left the ground at Indianapolis. The captain is probably making a joke or he is mistaken, since 3:15 Eastern Standard Time would be 2:15 Chicago time.

FLIGHT ATTENDANT: [*To crew*] What do you do up here when [you are] autopiloting? Just hang out?
COPILOT: You still gotta tell it [the autopilot] what to do.
CAPTAIN: If the autopilot did not work, he'd [the copilot, who was flying the aircraft, would] be one busy little bee right now.
CABIN: [*Laughter*]
FLIGHT ATTENDANT: [*To captain*] So, does the [copilot] do a lot more work than you do?
CAPTAIN: Yep.
CABIN: [*Laughter*]
COPILOT: Not really.
CAPTAIN: Man, this thing [the airplane] gets a high deck angle in these turns [in the holding pattern].
COPILOT: Yeah.
CAPTAIN: We're just wallowing in the air just now.
COPILOT: You want flaps fifteen?
CAPTAIN: [*Jokingly*] I'll be ready for that stall procedure here pretty soon [if the sluggishness of the airplane continues].
CABIN: [*Laughter*]
CAPTAIN: Do you want [to] kick 'em [the flaps] in? [It'll] bring the nose down.
COPILOT: Sure.

The flight attendant asks about the recorded warning voices in the cockpit. She thinks that they also warn about the presence of rain.

CAPTAIN: Rain?
COPILOT: No, this one maybe?
FLIGHT ATTENDANT: Sounds like it said something about the rain, or . . .
CABIN: [*The captain causes the alarm warning voices to sound for the flight attendant. Sound of warning: "Glide slope! Whoop, whoop! Pull up! Whoop, whoop! Pull up!"*]
CAPTAIN: That one [you mean]?
FLIGHT ATTENDANT: Yeah, but there's something else.
COPILOT: [*Joking with flight attendant*] Like I said, [if] it's a rain

cloud, [recorded voices] say [it's rain]. Well, how do you know [there is rain]? Because this thing tells me. It'll [ground-proximity warning system will] tell you: ter*rain*, ter*rain*.

FLIGHT ATTENDANT: That's what it says? Ter*rain*?

The captain or the copilot then plays the recorded voice for the flight attendant to hear.

CABIN: [*Sound of warning: "Too low terrain, too low terrain."*]
CABIN: [*Sound of a radio playing music*]
CABIN: [*Low-frequency sound starts and increases as the revolutions per minute of the propellers automatically increase*]
FLIGHT ATTENDANT: See you all.
CAPTAIN: All right.
CABIN: [*Sound of cockpit door being opened and closed*]
CABIN: [*Low-frequency sound decreases, indicating a decrease in propeller rpms*]
COPILOT: [*To captain*] Let's see. We got about, uh, thirty-six hundred pounds of fuel?
CAPTAIN: Uh-huh.
CABIN: [*Music stops*]
CAPTAIN: [*On public-address system to passengers*] Well, folks, once again this is the captain. [We] do regret to inform you that Air-Traffic Control has put us into a holding pattern up here. We're holding for approximately twenty minutes out of Chicago at this time, but, uh, I guess the congestion and traffic's continued on, uh. . . . They need us to hold out here for some spacing. They're saying at this point, uh, on the hour before we depart the hold, though that may not hold, uh. . . . We may not be here the full thirteen minutes. We'll be sure to keep you updated. Once we leave the hold, we'll let you know more if they tell us the hold is going to be a little bit longer. I do apologize for all these delays. Chances are that all the flights in and out of Chicago here this afternoon are going to be delayed as well. This is not just aircraft in the air just now but this is also, uh, for aircraft that were in the air earlier, aircraft on the ground, and aircraft that are going to be departing. So, uh, once again, chances are that your [connecting] flight would be delayed also, and you'll still have a real good chance of making your connection. If not, they'll, uh, automatically rebook you on the next flight in Chicago.

The captain and the copilot discuss a bit of shared confusion about the workings of ACARS, which is a message system from the ground, usually the company offices, to the cockpit, printed out on hard copy.

COPILOT: That's much nicer, flaps fifteen.
CAPTAIN: Yeah.
COPILOT: I'm showing some ice now.

The captain needs to go to the bathroom badly.

CAPTAIN: I can't hold [it] anymore, man. That big [thing?] needs out. Right now.
COPILOT: [*Chuckles*] They're gonna be giving you dirty looks [for going back through the cabin to the bathroom], man.
CAPTAIN: Oh, man—oh, yeah. I know they are. People [passengers] do. It's either that or pee on 'em.
CABIN: [*Sound of chime, like flight-attendant call bell; sound of seat belt being unfastened*]
CAPTAIN: [*To flight attendant on interphone*] What's up?
FLIGHT ATTENDANT: [*Calling the cockpit*] It's just me.
CAPTAIN: Huh?
FLIGHT ATTENDANT: I'm, uh . . . It was just me.
CAPTAIN: Oh.
FLIGHT ATTENDANT: I'm just wondering how much gas we got?
CAPTAIN: How much gas we got?
FLIGHT ATTENDANT: Yeah.
CAPTAIN: We got more than plenty of gas. We can be out here for a long time.
FLIGHT ATTENDANT: Cool. Okay. [I] just was worried [that] maybe you'd have to divert somewhere [for gas] and really make these people late.
COPILOT: Sixty miles from Chicago.
CAPTAIN: Oh, yeah.
FLIGHT ATTENDANT: Six, sixty miles?
COPILOT: Yeah.
FLIGHT ATTENDANT: Yeah, but they're still gonna hold us, huh?
CAPTAIN: Till about another ten minutes.
FLIGHT ATTENDANT: And that's not a for-sure thing, is it?

CAPTAIN: Yeah, pretty for-sure as of right now, unless they decide to make it different. How's that for an answer?
FLIGHT ATTENDANT: [*Chuckles*] Same like the other one.
CAPTAIN: Yeah, I know.
CAPTAIN: [*To copilot*] Talk to her, bro.

Here the captain gets up and goes back to the bathroom, leaving the copilot alone at the controls.

COPILOT: Okay. Hey, you there?
FLIGHT ATTENDANT: [*To copilot*] Are you sure you can handle it up there?
COPILOT: I'll try.
FLIGHT ATTENDANT: Okay, uh . . .
COPILOT: Why do you—
FLIGHT ATTENDANT: Turn it [the heat] down. It needs to be cooler back here. It's hot.
COPILOT: I'm, uh, it's all the way [turned] down now.
FLIGHT ATTENDANT: Okay, thanks. It's been down?
COPILOT: Yeah, well, I'll chill it up. . . .
FLIGHT ATTENDANT: Really? Well, we're sweatin'. [*Sound of panting*]
COPILOT: You know why?
FLIGHT ATTENDANT: You want to hear us breathe heavy? [*Laughs*]
COPILOT: It's one of the bleeds [that] are off.
FLIGHT ATTENDANT: Okay.
COPILOT: [One of the bleeds] for the air conditioning.
FLIGHT ATTENDANT: Yeah.
COPILOT: And it's [on] your side.
FLIGHT ATTENDANT: Oh.
COPILOT: It's the one that gives you most of the air back there.
FLIGHT ATTENDANT: Figures.
COPILOT: So now you got, you got less than, uh, half. Not only that. It's your half. [*Chuckles*]
FLIGHT ATTENDANT: Okay. Okay, well, here. Orlando [the captain] wants to talk to you.
COPILOT: Orlando does? Hello.
CAPTAIN: [*On interphone from the back of the airplane*] Hey, bro.
COPILOT: Yeah.
CAPTAIN: [I'm] gettin' busy with the ladies back here.
COPILOT: Oh.

FLIGHT ATTENDANT: [*Snickers*]
CAPTAIN: Yeah, so if, so if I don't make it up there within the next, say, fifteen or twenty minutes, you know why.
COPILOT: Okay. I'll, uh . . . When we get close to touchdown, I'll give you a ring.
CAPTAIN: There you go. No, I'll be up right now. There's somebody in the bathroom, so . . . Talk to you later.
COPILOT: Okay.
CAPTAIN: [*Reentering the cockpit*] We have a whole new hombre [after going to the bathroom].
CAPTAIN: Did you get any more messages from the cabbage patch [from the company on ACARS]?
COPILOT: No . . .

A minute goes by. The crew discusses where they put the paper with the list of connecting gates at O'Hare.

CAPTAIN: We still got ice.
CABIN: [*Sound of paper being torn from the ACARS printer*]
CAPTAIN: Here.
COPILOT: Got a message?
CAPTAIN: You did.
CAPTAIN: I'll be right back. Okay, I'm [going] to talk to the company.
CENTER: Eagle Flight One eighty-four, descend and maintain eight thousand [feet].
CAPTAIN: [*Calling the company offices at O'Hare*] Chicago, do you copy, Forty-one eight four?
COMPANY: Forty-one eighty-four. Go ahead.
CAPTAIN: Yeah, we've already been talking to Dispatch, uh, on the ACARS, so they are aware of our delay. I don't know if you guys got the word on that. We're on a hold out here, uh. . . . We got thirty-two hundred pounds [of fuel], thirty-three hundred pounds of fuel. They're saying zero zero . . . so in about another four or five minutes we'll find what the new word is. But what can you tell me about, um . . . There's this guy [passenger who is] concerned about his Frankfurt [Germany] connection. Uh, do you know anything about that?
CENTER: Eagle Flight One eighty-four, descend and maintain eight thousand [feet].
COPILOT: Down to eight thousand, Eagle Flight One eighty-four.

CENTER: Eagle Flight One eighty-four, uh, should be about ten minutes, uh, till you're cleared in.

COMPANY: Uh, I can double-check on that, uh, yeah. Just sent a message to Dispatch to see if you were in a hold. Copy thirty on the fuel and estimated out time on the hour. And did you have that, uh, Frankfurt flight number, by any chance?

COPILOT: They say ten more minutes [until cleared to land].

CAPTAIN: Um, no. I sure don't, but I pulled up connecting gates out of the ACARS, and it says it's [the Frankfurt flight is] going out of [Gate] K-Five, if that helps at all.

COMPANY: Let me check.

CAPTAIN: [*To copilot*] Are we out of the hold [yet]?

COPILOT: Uh, no. We're just goin' to eight thousand.

CAPTAIN: Okay.

COPILOT: And, uh, ten more minutes, she [the controller] said.

CABIN: [*Sound of beeps warning of overspeed*]

COPILOT: Oops.

CAPTAIN: We . . . I knew we'd do that.

COPILOT: I's tryin' to keep it at one eighty [mph].

CABIN: [*Repetitive thudding sound*]

CABIN: [*Wailing sound warning of pitch-trim movement*]

It is at this point that, without any warning, the airplane flies out of control. It rolls rapidly to the right. In a matter of seconds, the airplane is falling at 500 feet per second, pulling a g-force rating of 3.6.

COPILOT: Oops.

CABIN: [*Sound of three thumps followed by rattling; sound of chirps consistent with the autopilot disconnecting; sound of altitude-alert signal*]

CREW: Okay. [*Intermittent heavy irregular breathing starts and continues to end of recording*]

CREW: Oh, shit.

CAPTAIN: Okay.

CABIN: [*Sound of altitude-alert horn*]

CREW: [*Sound of a growl continues for next twelve to thirteen seconds, until impact*]

CAPTAIN: All right, man. Okay, mellow it out.

COPILOT: Okay.

CAPTAIN: Mellow it out.
COPILOT: Okay.
CAPTAIN: Nice and easy.
CABIN: [*Terrain warning: "Whoop, whoop! Terrain."*]
COPILOT: Aw, shit.
CABIN: [*Sound of loud crunching*]

For the next sixteen minutes, flight controllers at Chicago-O'Hare ask seventeen times for Flight 4184 to respond.

END OF TAPE

All sixty-four passengers and the crew of four died instantly when Flight 4184 crashed nose-first in a wet soybean field sixty miles from Chicago. The NTSB concluded that the probable cause of the accident was "the loss of control, attributed to a sudden and unexpected aileron hinge movement reversal that occurred after a ridge of ice accreted beyond the deice boots. . . ." In short, ice interfered with the proper operation of the control surfaces. The autopilot held the airplane in level flight until it could hold it no more. Then it switched off automatically, handing over the controls of an out-of-control airplane to the flight crew.

Dallas-Fort Worth, Texas

August 31, 1988

■

Delta Air Lines Flight 1141

This incident typifies the adage that if something can go wrong it will, sometimes tragically, as in the case of Delta Flight 1141. While still on the ground at the gate at Dallas–Fort Worth International Airport, the Boeing 727-232 was serviced and prepared for a flight that would take it to Salt Lake City, Utah. The weather that afternoon, according to the captain of the flight, was "beautiful."

On board were ninety-eight passengers, three flight-crew members, and four cabin attendants.

Before the engines were started, even before the captain came aboard the airplane, a fueling agent checked the fuel in the number-one fuel tank, then, according to habit, measured the airplane's pitch and roll (its overall balance) by the use of a plumb line in the aircraft's right main rear wheel well. Together, these readings were radioed to Delta Operations' senior customer service agent, who is responsible for the proper loading of fuel on the aircraft. The agent determined that the aircraft needed 1,036 gallons in the number-one tank for a total of 1,597 gallons. An equal amount was ordered as a total for the number-three tank. The number-two tank was ordered to be filled to 10,600 pounds final weight.

The fueling agent went aboard the aircraft when he was finished to check the fuel gauges in the cockpit. The fuel gauge for number-two tank read 500 pounds higher than when he was on the ground fueling the aircraft. He mentioned this discrepancy to the copilot.

The aircraft had landed in Dallas–Fort Worth with a number-one fuel-quantity gauge in the cockpit that did not work. The copilot knew about the gauge, as did the Delta mechanic who had handled the aircraft while it was on the ground. The gauge was not fixed, but its status was noted in the aircraft's logbook.

While the flight was taxiing out for takeoff at one-thirty that afternoon, a lengthy delay inspired the captain to shut down the number-three engine to conserve fuel. When the flight was number three in line for takeoff, the engine was restarted. The delay lasted approximately thirty minutes.

Flight 1141 was already an accident about to happen.

We pick up the CVR as they are taxiing out for takeoff.

DALLAS–FORT WORTH GROUND CONTROL: Eleven forty-one, give me a right turn. Bring it between [the] south ramp and [Taxiway] Thirty. Hold short of inner [taxiway].

COPILOT: Eleven forty-one, roger.

Two minutes go by.

CAPTAIN: [*Referring to the lack of attention given to them by Ground Control*] How about lookin' down here at Delta, now and then?

COPILOT: Yeah.

CABIN: [*Laughter*]

FLIGHT ENGINEER: [Pay attention to us] while we're still young [referring to the delay]. How about lookin' down here while we still have teeth in our mouths?

CAPTAIN: [*To flight engineer*] What's that?

FLIGHT ENGINEER: [I said] how about lookin' down our way while we still have teeth in our mouths.

CABIN: [*Laughter*]

COPILOT: [We're] growing gray at the south ramp. . . .

CAPTAIN: [The delay is] too long. I guess we ought to shut down number three [engine] to save a few thousand dollars [on fuel].

COPILOT: I'll, I'll call and ask Ground [Control] if we can . . . just, like, shut down over here. [I'll] ask him if he can give us a two-minute warning to start our engines.

FLIGHT ENGINEER: Okay.

GROUND CONTROL: . . . Make a left turn and say your [flight] number.

COPILOT: That's Eleven forty-one.

GROUND CONTROL: Delta Eleven forty-one, okay. Give way to company [another Delta jet] to your left, the [Boeing] seven two [seven]. Join the inner for standard taxi [for takeoff on] Eighteen-L.
COPILOT: Ah, seven-thirty, roger.
CAPTAIN: We're gonna wait for him [Delta's Boeing 727]?
COPILOT: Yeah.
CAPTAIN: Where is it [Delta Boeing 727]?
COPILOT: He's right there.
FLIGHT ENGINEER: He's comin' out.
CAPTAIN: [We] certainly taxied out before he did. Did he say standard to [Runway] Eighteen?
FLIGHT ENGINEER: Takeoff data has been computed for Eighteen-L.
COPILOT: Okay.

For the next fifteen seconds, the crew goes through the pre-takeoff checklist.

FLIGHT ATTENDANT: A lotta people goin' out this morning.
FLIGHT ENGINEER: Yeah, [a] big blitz.

Now there are nearly eight minutes of conversation between the crew and the flight attendant.

CAPTAIN: Don't we have to change to ground here?
COPILOT: Yeah, I'm sorry. I'm sittin' here talking to the flight attendant.

The flight is given further taxiing instructions from Ground Control.

FLIGHT ATTENDANT: Are we going to take off or are we just gonna roll around the airport?
COPILOT: Well, we thought we were going to have to retire, sittin' here. They're [Delta's Boeing 727 is] waiting for taxi clearance.
FLIGHT ATTENDANT: My gosh, we have got a long taxi to do.
CAPTAIN: Yeah, we're gettin' down here where we let all the Americans [American Airlines flights in line for takeoff] get off first. Once they are all gone, we can go.

Another minute and a half go by.

COPILOT: What kind of birds are those?
FLIGHT ENGINEER: Egrets, or whatever they call them.

FLIGHT ATTENDANT: Yeah, egrets.

COPILOT: Are they?

FLIGHT ATTENDANT: I think so. Are they a cousin to the ones by the sea?

CAPTAIN: I don't know. . . . Whenever I mow the grass out in my pasture they come in and it stirs up the grasshoppers and everything.

FLIGHT ENGINEER: Boy, they just flock here.

COPILOT: I've seen them all over the place out around here.

CAPTAIN: Grasshoppers . . . They [egrets], in fact, they sit on the back of our horse now and then. You see one out there just sittin' on the back of the horse.

FLIGHT ATTENDANT: Oh, is that right?

COPILOT: I've seen them sittin' on the back of a lot of cows.

CAPTAIN: Yeah.

FLIGHT ATTENDANT: Are they the ones that pick the bugs off of them and stuff?

CAPTAIN: I guess, and hang around them, because while they're [cows are] grazing, you know, they stir up the insects, and they [the egrets] can get them easier.

FLIGHT ATTENDANT: Uh-huh. They're pretty birds.

COPILOT: They got one more American [Airlines flight in line] and I think we'll be able to go here. Start clearin' some of this. . . .

CAPTAIN: It's interesting how they [the egrets] sit around the airport like this without being afraid.

SECOND FLIGHT ATTENDANT: Didn't we taxi [yet]?

COPILOT: I'm surprised they're [the egrets are] not complainin' about the [jet] noise.

FLIGHT ATTENDANT: [*To second flight attendant*] We go to the other side [of the runway].

CAPTAIN: [*To flight attendant*] All the way around down here, back up here.

FLIGHT ENGINEER: Boy, this [delay] is somethin'. If this is how it is on a severe clear VFR [a beautiful] day, can you imagine . . .

COPILOT: . . . Imagine what it'd be like if we had weather today. I tell you what. Dallas–Fort Worth Center is a hundred percent better than it was three years ago.

FLIGHT ATTENDANT: Is it really?

COPILOT: With regard to weather—gettin' in and getting out and all. If it clouded up and even looked like there was going to be weather . . . you'd be holdin', slowin' down, all kinds of stuff.

CAPTAIN: Did you see that bird?

FLIGHT ATTENDANT: Yeah.

COPILOT: He just got jet blast.

FLIGHT ATTENDANT: Yeah, he did. He got it.

COPILOT: Ah, what a crash.

FLIGHT ATTENDANT: [The egret must have said] what in the world was that?

COPILOT: [Did you] ever go out to Midway [Island in the South Pacific] and see the gooney birds? They're somethin' to watch.

FLIGHT ENGINEER: [The gooney birds] crash and look around to see if anybody saw 'em, you know. . . .

COPILOT: Yeah. They would. . . . If you would do a [engine] runup, and the gooney birds would be back there in the prop[eller] wash just hangin' in the air, you know. And then they'd pull the power back [on the engines] and then [the gooney birds] would just . . .

CABIN: [*Laughter*]

COPILOT: . . . Hit the ground, you know. They were hilarious. They'd send a truck out. You'd get ready to take off. They'd send a pickup truck out and they'd go move the birds off the runway so you could take off.

FLIGHT ATTENDANT: Oh, really? Oh, how funny. Where are they? Where was that?

COPILOT: Midway, Midway Island. They come back and they nest in exactly the same spot [where] they were born.

FLIGHT ATTENDANT: On the runway?

COPILOT: Yeah, whether it was a runway or what[ever] it was. They come back to the exact same spot and, ah, so there's some kinda law or somethin' that you can't build anything on the island anymore because . . .

FLIGHT ATTENDANT: Uh-huh . . .

COPILOT: It's [the island is] a sanctuary for the birds or somethin'.

FLIGHT ENGINEER: [*On public-address system to the main cabin*] Good morning, ladies and gentlemen. We're number four for departure. Flight attendants, prepare the cabin, please.

FLIGHT ATTENDANT: We're ready [to go].

FLIGHT ENGINEER: Thank you.

CABIN: [*Cockpit door is closed*]
CAPTAIN: Might as well start [number-three engine].
COPILOT: Number three.
FLIGHT ENGINEER: Start valve open.
CABIN: [*Sound of engine starting*]

Thirty seconds go by.

TOWER: Eleven forty-one, taxi position [to] Runway Eighteen left and hold. . . .
FLIGHT ENGINEER: Shoulder harness.
COPILOT: They're on.
FLIGHT ENGINEER: Flaps.
COPILOT: Fifteen. Fifteen. Green light.
FLIGHT ENGINEER: Flight controls.

For the next fifteen seconds, the crew goes through the pre-takeoff checklist.

TOWER: Delta Eleven forty-one, fly heading one eight five, Runway Eighteen left, cleared for takeoff.
COPILOT: Eleven forty-one, one one eight five, cleared for takeoff.

The brakes are released, power is applied, and the aircraft begins its takeoff roll down the runway, picking up speed rapidly.

CABIN: [*Sound of engines spooling up*]
COPILOT: Power is set. Engine instruments look good. Airspeed is comin' up both sides. Eighty knots. Vee R, Vee two.
CABIN: [*Sound of snap; sound of the stick shaker*]

Without warning, the aircraft rolls violently to the right.

CAPTAIN: Somethin's wrong! Oh . . .
CABIN: [*Sound of engine stall*]
COPILOT: Engine failure.
CAPTAIN: We got an engine failure. We're not gonna make it. Full power.
CABIN: [*Sound of first impact; sound of second impact; sound of third impact*]
CABIN: [*Screams*]
CABIN: [*Sound of fourth impact*]

END OF TAPE

The captain later stated that "all was normal" during the takeoff roll, 6,017 feet down the runway at 177 mph. As the main gear wheels left the ground, the airplane began to roll violently, causing the right wing tip to contact the runway at 1,033 feet after liftoff. "It was all I could do to control the plane," the captain said. He called for full power. He continued to raise the plane's nose. The captain said, "We're not going to make it." There was a "horrendous impact." The aircraft struck an antenna array on the ground 1,000 feet beyond the end of the runway, and it came to rest 3,200 feet beyond the departure end of the runway. The airplane was destroyed by the impact and a post-crash fire. Fourteen people died.

The cause of the crash was later determined. The takeoff "configuration" was wrong for the aircraft. Its weight distribution, principally the fuel in the tanks, caused the airplane to roll out of control the instant it left the ground.

Cape Canaveral, Florida

January 28, 1986

■

Space Shuttle *Challenger*

The *Challenger* space shuttle was hardly a commercial airplane, or even an airplane at all. But it did contain a black box that recorded the voices and sounds in the cockpit during the seconds before launch and the few seconds in flight before the rocket destroyed itself over the Atlantic in a catastrophic failure of its booster-engine systems.

The commander of the shuttle that cool morning was Francis Richard Scobee, the pilot was Michael J. Smith, and the mission specialists were Ellison S. Onizuka and Judith A. Resnik.

The tape runs for the period of T-2:05 (two minutes and five seconds) prior to launch, when the CVR automatically activated, through T+73 seconds after launch, until LOSS OF DATA.

Mission specialist Ronald E. McNair and payload specialists S. Christa McAuliffe and Gregory B. Jarvis, who were seated for launch in *Challenger*'s middeck, were out of range of the cockpit voice recorder's microphones.

T-2:05: RESNIK: [*Probably speaking to Commander Scobee*] Would you give that back to me? [It's my] security blanket. Hmm.

T-1:58: SCOBEE: Two minutes, downstairs; you got a watch running down there?

There were now two minutes to launch.

T-1:47: SMITH: Okay, there goes the LOX [liquid oxygen supply] arm.
T-1:46: SCOBEE: [There] goes the beanie cap [liquid oxygen vent cap].
T-1:44: ONIZUKA: Doesn't it go the other way?
T-1:42: CABIN: [*Laughter*]
T-1:39: SCOBEE: Now I see it. I see it.
T-1:39: SMITH: God, I hope not, Ellison [Onizuka].
T-1:38: SCOBEE: I couldn't see it [liquid oxygen supply arm] moving; it was behind the center screen.
T-1:33: RESNIK: Got your harnesses [seat restraints] locked?
T-1:29: SMITH: [*Joking*] What for?
T-1:28: SCOBEE: I won't lock mine; I might have to reach something.
T-1:24: SMITH: Ooh-kaaay.
T-1:04: ONIZUKA: Dick [Scobee]'s thinking of somebody there.
T-1:03: SCOBEE: Un-huh. One minute, downstairs.

One minute to launch.

T-52: RESNIK: Cabin pressure is probably going to give us an alarm [a caution and warning alarm that was routine during prelaunch].
T-50: SCOBEE: Okay. Okay there.
T-43: SMITH: Alarm looks good [cabin pressure is acceptable].
T-42: SCOBEE: Okay.
T-40: SMITH: Ullage pressures are up. Right-engine helium tank is just a little bit low.
T-32: SCOBEE: It was yesterday, too.
T-31: SMITH: Okay.
T-30: SCOBEE: Thirty seconds down there.

Thirty seconds to launch.

T-25: SMITH: Remember the red button when you make a roll call [reminder for communications].
T-23: SCOBEE: I won't do that; thanks a lot. Fifteen [seconds to launch].
T-6: SCOBEE: There they go, guys.

Ignition of boosters.

Resnik: All right.

Scobee: Three at a hundred [thrust level at 100 percent for all three engines].

T+0: Resnik: Aaall riiight.

T+1: Smith: Here we go [vehicle motion].

T+7: Scobee: Houston, *Challenger* roll program [initiation of vehicle roll program].

T+11: Smith: Go, you mother.

T+14: Probably Resnik: LVLH [reminder for cockpit-switch configuration change].

T+15: Resnik: [Expletive] hot.

T+16: Scobee: Ooh-kaaay.

T+19: Smith: Looks like we've got a lotta wind here today.

T+20: Scobee: Yeah. It's a little hard to see out my window here.

T+28: Smith: There's ten thousand feet and Mach point five [altitude and velocity report].

T+35: Scobee: Point nine [velocity report, 0.9 Mach].

T+40: Smith: There's Mach one [velocity report, 1.0 Mach].

T+41: Scobee: Going through nineteen thousand [feet]. Okay, we're throttling down [normal thrust reduction].

T+57: Scobee: Throttling up [throttle up to 104 percent after maximum dynamic pressure].

T+58: Smith: Throttle up.

T+59: Scobee: Roger.

T+60: Smith: Feel that mother go. Woooo-hoooo. Thirty-five thousand [feet] going through one point five [altitude and velocity report, 35,000 feet, 1.5 Mach].

T+1:05: Scobee: Reading four eighty-six on mine [routine airspeed-indicator check].

T+1:07: Smith: Yep, that's what I've got, too.

T+1:10: Scobee: Roger, go at throttle up [boosters at 104 percent].

T+1:13: Smith: Uh-oh.

T+1:13: LOSS OF ALL DATA.

END OF TAPE

Cali, Colombia

December 20, 1995

■

American Airlines Flight 965

The night of December 20 was starry over the Andes. Aboard Flight 965, an American Airlines Boeing 757 on a regularly scheduled passenger flight from Miami to Alfonso Bonilla Aragón International Airport in Cali, Colombia, there were 155 passengers flying home for the Christmas holidays, as well as 6 cabin attendants and 2 cockpit-crew members looking forward to a break in their schedule. The flight had left Miami late, because of late-arriving connecting passengers and a one-hour-and-twenty-one-minute ground delay caused by airport traffic. But after takeoff, Flight 965 ascended to 37,000 feet over Cuban, Jamaican, and into Colombian air space.

There was no evidence of a malfunction in the airplane. The weather was beautiful, but of course it was night over the Andes.

We start the CVR transcript with the flight a few minutes away from its initial descent into Cali over the caldera of the northern Andes. The aircraft was cruising at 37,000 feet with the center autopilot and both flight directors engaged. Heading was 189 degrees. The captain had apparently left the cockpit for the bathroom, and he returned, asking the copilot what had happened in his absence. They discussed the issues of crew layovers and scheduling.

COPILOT: Well, we did get the weather. It's good.
CAPTAIN: All right. She [the flight attendant in the back, where he had just come from] is claiming they get an extra twenty minutes.
COPILOT: An extra twenty minutes for what?
CAPTAIN: It's difficult [to explain] with the language problem,

but . . . according to her figures, they're not legal to report to the airport until eight-fifty. If we get in [land] at ten o'clock . . .

For the next several seconds, the captain outlines the problems of his cabin crew; according to FAA regulations they must not fly more than a set number of hours without a rest, and those hours for this cabin crew are nearly up.

COPILOT: I got this little chart [that tells the FAA regulations about rest versus flying].

CAPTAIN: Well, you see what you come up with. I'll watch the airplane and the radio. Okay?

COPILOT: Okay. All I see on this little chart they handed out is duty-on time, but it doesn't say anything about rest period.

CAPTAIN: That's another very confusing thing. . . . I started to say, I wrote this little sheet out. I called [crew] tracking one day and I said, "Hey, this fucking international [flying] is doing me, and I don't understand how a two-man crew . . . blah, blah, blah." I said, "I want you to spell out the legal rest," and that's where I got this [little chart] from. And I wrote it down very explicitly. Ten hours minimum crew rest.

COPILOT: That's on international [flights]?

CAPTAIN: Yeah, if you fly less than five and a half hours.

COPILOT: Which is the case [here].

CAPTAIN: That's our scenario. Ten hours crew rest, thirty minutes debrief, and one hour sign-in. And you can't move that up [change that] at all, because it's an FAA thing. You roll those wheels before eleven and a half hours, and you're fucked. Now, now, like I say. I can't, I'll have you know, grab a little extra half hour for us. We'll report a little bit late, just [to] give us a little extra sleep[ing] time. As long as we get the thing [the airplane] off at nine-fifty [the next morning], so we don't get our asses [caught by the clock], [and they'll ask] why the hell didn't you report? To which I will say, "The thirty- [or] forty-minute cab ride each way [from the airport to the hotel and] I don't think we had enough legal safe time. Now, if you want to hang me on that, you hang me on that, but I didn't break any FAA regulations. Anyway, you know. . . . When you want descent, let me know a few minutes early in case there's a language problem [with the local controllers], okay?

COPILOT: Sure.
CAPTAIN: I can get through [to them linguistically].

Again, they complain about cockpit fatigue, mentioning a "miserable last four hours" and "these eight-and-a-half-hour deals."

CAPTAIN: Yeah, a friend of mine—I played tennis with him, and, uh, he used to fly that São Paulo [Brazil route] all of that time—you're fuckin' killin' yourself doin' that shit. [Do] you really need that extra couple hundred bucks a month or whatever when it comes to retirement? But anyway, to each his own. But he said he didn't mind it. He didn't mind driving back home at five o'clock in the morning. But to me, I'm like . . . it's torture.
COPILOT: Yeah.
CAPTAIN: Torture in the car trying to keep awake and stay alive. I discussed this with my wife. I said, "Honey, I just don't want to do this, [and] I hope you don't feel like I'm [unintelligible words]." She said, "No way, forget it." She said, "You don't need to do that shit." [*Sound of a yawn*] Yeah, he [the pilot who didn't mind driving at five in the morning] just retired [from the airlines] a couple weeks ago.
COPILOT: Yeah, I knew this was his last month.
CAPTAIN: Yeah, he's a good man. I like him. We're good friends.
COPILOT: He got robbed at knifepoint in Rio, wasn't it?
CAPTAIN: That's right. He got stuck [by the knife] a little bit, actually, too.
COPILOT: [*Turning to the business of flying*] Well, let's see. We got a hundred and thirty-six miles to the VOR, and [we have] thirty-two thousand feet to lose, and slow down to boot, so we might as well get started [with our initial descent].
CAPTAIN: All right, sir. And if you'd keep the speed up in the descent, it would help us, too, okay?
COPILOT: Okay.
CAPTAIN: [*To Bogotá Control*] Bogotá, American Nine sixty-five, request descent.
BOGOTÁ CONTROL: American Nine sixty-five, descend and maintain flight level two four zero [24,000 feet], report reaching [that altitude].
CAPTAIN: Okay, we're leaving three seven zero [37,000 feet]. Descend and maintain two four zero, twenty-four [24,000 feet]. Thank you, ma'am. American Nine sixty-five.

CONTROL: That's correct.
COPILOT: Twenty-four [thousand feet] set.
CAPTAIN: Yes, sir. I'm goin' to call the company.
COPILOT: Okay.
CAPTAIN: American Airlines Operations at Cali, this is American Nine sixty-five, do you read?
OPERATIONS: Go ahead, American Nine six five, this is Cali Ops.
CAPTAIN: All right, Cali. We will be there in just about twenty-five minutes from now. Eh, and go ahead [and give us] the weather.
OPS: Okay, sir . . . The temperature is twenty . . . degrees. The altimeter is two nine point eight, conversion to two six point seven one.
CAPTAIN: Okay, understand the weather is good. Twenty-three degrees, to nine nine eight. Two six seven one. Is that correct?
OPS: That's correct.
CAPTAIN: Okay, are we parking at Gate Two tonight?
OPS: Gate Two and, uh, Runway Zero one.
CAPTAIN: Runway Zero one, and the weather is good, huh?
OPS: Okay, captain.
CAPTAIN: See you on the ground, Nine six five.

The captain then repeats the information on weather, heading, and wind and the runway and gate numbers to the copilot, who is flying the airplane.

COPILOT: Okay.
CAPTAIN: All right, baby.
COPILOT: Sounds good.
CAPTAIN: All right. And I'm gonna put the headlights on early here because there's a lot of VFR [visual flight traffic in the area] and who knows what good deal. So the headlights just might help us a little bit [to let other aircraft see us]. And also, what was that position? Was [it] five? We're just about at it, aren't we?
COPILOT: Yeah. Forty-seven [miles] north of Río Negro. Uh, of course we didn't go to Río Negro.
CAPTAIN: Sorry?
COPILOT: [I was] talking about the, uh . . .
CAPTAIN: Yeah, it was Río Negro plus forty-seven, I think. . . .

COPILOT: Río Negro plus forty-seven.

CAPTAIN: . . . What's, what [does the flight plan] show lat-long [for latitude and longitude]?

COPILOT: [*Looks for flight plan*] Well, let me find it.

CAPTAIN: Just out of curiosity.

COPILOT: I had the flight plan. . . .

CAPTAIN: All right, I wouldn't worry about it.

COPILOT: [*Finds the flight plan*] There we go. North, uh, zero five one four six four, so zero five forty-one . . .

CAPTAIN: We're past it. Okay, we're past it. We press on, right?

COPILOT: Right.

CAPTAIN: I'm going to talk to the people [passengers].

COPILOT: Okay.

CAPTAIN: I'm off. [*Now into public-address system*] Uh, ladies and gentlemen, this is [the] captain. We have begun our descent for landing at Cali. It's a lovely evening as we had expected. We'll pass a shower or two on the way in, but, uh, at the field right now it's, uh, good visibility, the temperature is two three, that's twenty-three degrees Celsius, and if you prefer Fahrenheit, that's seventy-two degrees on the Fahrenheit scale. The winds are ten miles an hour from the northwest. It's a very, very pretty evening. I'd like to thank everyone for coming with us. Again, I apologize for being late tonight. These things do happen sometimes, very frustrating, but there wasn't very much we could do about it. Again, I appreciate your patience in the matter. Like to wish everyone a very, very happy holiday, and a healthy and prosperous 1996. Thank you for coming with us. [*To copilot*] I'm back.

COPILOT: Uh, I may have to slow down if it gets too rough.

CAPTAIN: Sure.

Now they are descending through 25,700 feet to their assigned level of 24,000 feet.

CAPTAIN: You want any of these nuts?

COPILOT: No, thank you.

CAPTAIN: You want me to call for the water or do you want to wait till we get on the ground, 'bout your water?

FIRST OFFICER: Oh, I'll get it on the ground. One to go.

CAPTAIN: Aye-aye. You got the engine heat off. Good. [*To Bogotá Control*] American Nine six five is [at] level two four zero [24,000 feet]. American Nine six five is level two four zero.

CONTROL: Stand by two minutes for [a] lower [altitude].

CAPTAIN: [*To copilot*] Pretty night, huh?

COPILOT: Yeah, it is lookin' nice out here. Let's see, what is the transition level here?

CAPTAIN: Oh, yeah, it's a good check. Eighteen thousand?

COPILOT: One ninety [mph], eighteen thousand, yeah. Well, if she [Bogotá controller] doesn't let us down in a little while, she's goin' to put me in a jam here.

CAPTAIN: [*To Control*] And American Nine six five, request lower [altitude].

CONTROL: American Nine six five, descend to flight level two zero zero [20,000 feet]. Report leaving two four zero.

CAPTAIN: We're leaving two four zero now and descending to two zero zero.

CONTROL: Call Cali frequency one one niner decimal one. *Buenas noches* [Good night].

CAPTAIN: Please say the frequency again.

CONTROL: One one niner decimal one.

CAPTAIN: One one niner decimal one. *Feliz Navidad, señorita* [Merry Christmas, miss].

CONTROL: *Muchas gracias, lo mismo* [Many thanks, and the same to you].

CAPTAIN: *Gracias*. Center, American Nine six five, leaving flight level two four zero, descending to two zero zero. *Buenas tardes* [Good afternoon].

COPILOT: Nineteen one or . . .

CAPTAIN: That's Cali. Cali Approach, American Nine six five.

CALI APPROACH: American Niner six five, good evening. Go ahead.

CAPTAIN: *Ah, bueno, señor*, American Nine six five leaving two three zero, descending to two zero zero. Go ahead, sir.

APPROACH: The, uh, distance DME from Cali?

CAPTAIN: The DME is six three.

APPROACH: Roger, is cleared to Cali VOR, uh, descend and maintain one five thousand [15,000] feet, altimeter three zero zero two. . . .

COPILOT: One five.

Now Flight 965 turns right from heading 189 to heading 198.

APPROACH: No delay expect for approach. Report [when you reach], uh, Tulua VOR.

CAPTAIN: Thank you. [*To copilot*] I put direct Cali for you in there [a reference to the airplane's Flight Management System computer, called the FMS].

COPILOT: Okay. Thank you . . . Two-fifty below ten here?

CAPTAIN: Yeah. [*On public-address system to cabin*] Uh, flight attendants, please prepare for landing, thank you. [*To copilot*] I sat 'em down and—

APPROACH: Niner six five, Cali.

CAPTAIN: Niner six five, go ahead, please.

APPROACH: Sir, the wind is calm. Are you able to approach Runway One niner [19]?

CAPTAIN: [*To copilot*] Would you like to shoot the One nine [Runway 19] straight in?

COPILOT: Uh, yeah. We'll have to scramble to get down. We can do it.

CAPTAIN: [*To Approach*] Uh, yes, sir. We'll need a lower altitude right away, though.

APPROACH: Roger. American Nine six five is cleared to VOR DME approach Runway One niner.

CAPTAIN: [We are] cleared the VOR DME to One nine, Rozo one arrival. Will report the VOR, thank you, sir.

APPROACH: Report, uh, Tulua VOR.

CAPTAIN: Report Tulua. I gotta give you to Tulua first of all. You, you wanna go right to Cali, or to Tulua?

COPILOT: Uh, I thought he said the Rozo one arrival?

CAPTAIN: Yeah, he did. We have time to pull that out. . . . [*He refers to his charts; sound of rustling pages*] And, Tulua one . . . Rozo . . . There it is. Yeah, see, that comes off Tulua.

Flight 965 is now passing 17,358 feet with speed brakes deployed full over the next forty-two seconds. Airspeed decreases from 347 mph.

CAPTAIN: Can American Airlines, uh, Nine six five go direct to Rozo and then do the Rozo arrival, sir?

Approach: Affirmative. Take the Rozo one and Runway One niner. The wind is calm.
Captain: All right, Rozo, the Rozo one to One nine, thank you, American Nine six five.
Approach: Thank you very much . . . Report Tulua and [unintelligible] twenty-one miles, ah, five thousand feet.
Captain: Okay, report Tulua twenty-one miles and five thousand feet, American Nine, uh, six five.
Copilot: Okay, so we're cleared down to five [thousand feet] now?
Captain: That's right, and . . . off Rozo . . . which I'll tune here. See what I get.
Copilot: Yeah.
Captain: At twenty-one miles at five thousand's part of the approach. Okay?
Copilot: Okay.

Now the captain inputs data into the computer.

Approach: American Niner six five, [what is your] distance now?
Captain: Uh-huh, what did you want, sir?
Approach: Distance DME.
Captain: Okay, the distance from, uh, Cali is, uh, thirty-eight [miles].
Copilot: Uh, where are we? We goin' out to . . .
Captain: Let's go right to, uh, Tulua, first of all. Okay?
Copilot: Yeah. Where [are] we headed?
Captain: . . . I don't know. . . . What's this . . . ? What happened here?
Copilot: Manual . . .
Captain: Let's come to the right a little bit.
Copilot: . . . Yeah, he's wantin' to know where we're headed.
Captain: . . . I'm goin' to give you direct Tulua. . . .
Copilot: Okay.
Captain: . . . [Turn] right now. Okay, you got it?
Copilot: Okay.
Captain: It's on your map. [Or it] should be.
Copilot: Yeah, it's a left, uh, left turn.

Flight 965 rolls from left turn to right turn as its altitude passes through 13,600 feet.

CAPTAIN: Yeah, I gotta identify that . . . though I . . . Okay, I'm gettin' it. Seventeen seven. [It] just doesn't look right on mine. I don't know why.

COPILOT: Left turn, so you want a left turn back around . . . ?

Flight 965 rolls from right turn to left turn, then into right turn again.

CAPTAIN: Naw . . . Hell, no, let's press on to—

COPILOT: Press on to where, though?

CAPTAIN: Tulua.

COPILOT: That's a right [turn].

CAPTAIN: Where [are] we goin'? Come to the right. Let's go to Cali. . . . We got fucked up here, didn't we?

COPILOT: Yeah.

CAPTAIN: . . . How did we get fucked up here? Come to the right, right now, come to the right, right now.

COPILOT: Yeah, we're, we're in a heading select to the right.

CAPTAIN: And American, uh, thirty-eight miles north of Cali, and you want us to go Tulua and then do the Rozo, uh, to, uh, the runway, right? To Runway One nine?

APPROACH: Niner six five, you can land, Runway One niner. You can use Runway One niner. What is your altitude and the DME from Cali?

CAPTAIN: Okay, we're thirty-seven [miles] DME at ten thousand feet. [*To copilot*] You're okay. You're in good shape now. We're headin' . . .

APPROACH: Report, uh, five thousand and, uh, final to one one, Runway One niner.

CAPTAIN: We're headin' the right direction, you wanna . . .

Flight 965 passes through 10,000 feet in mountainous terrain to 5,000 feet. Neither crew member makes an attempt to stop the descent despite knowing that they have gone off the published approach course for landing at Cali and are in a valley surrounded by mountains. One minute from crashing, the copilot, who recognizes the deviation in their course, attempts to turn back to the "extended center line" of the runway, which is Rozo.

CAPTAIN: Come to the right, come come right to Ca— Cali for now, okay?

COPILOT: Okay.
CAPTAIN: It's that fuckin' Tulua I'm not getting for some reason. See, I can't get [it]. Okay now, no, Tulua's [fucked] up.
COPILOT: Okay. Yeah. But I can put it in the box if you want it. I don't want Tulua. Let's just go to the extended center line of, uh . . .
CAPTAIN: Which is Rozo? Why don't you just go direct to Rozo, then. All right?
COPILOT: Okay, let's . . .
CAPTAIN: I'm goin' to put that over you.
COPILOT: . . . Get some altimeters, we're out of, uh, ten [thousand feet] now.
CAPTAIN: All right.
APPROACH: Niner six five, [report your] altitude?
CAPTAIN: Nine six five, nine thousand feet.

Now they pass through 8,600 feet.

APPROACH: Roger, distance now?
CABIN: [*"Terrain, terrain, whoop, whoop."*]

Autopilot disengages. The flight crew adds full power and raises the nose of the aircraft; the speed brakes, which had been deployed, are not retracted.

CAPTAIN: Oh, shit . . . Pull up, baby.
CABIN: [*Sound of aircraft stick shaker*]

Flight 965 rolls out of right turn. A radar-altitude alert sounds. Landing gear and flaps are up.

COPILOT: It's okay.
CAPTAIN: Pull up. Okay, easy does it, easy does it.
CABIN: [*Sound of autopilot-disconnect warning; sound of aircraft stick shaker stops*]
COPILOT: Nope.
CAPTAIN: Up, baby . . .
CABIN: [*Sound of aircraft stick shaker starts and continues to impact*]
CAPTAIN: . . . More, more.
COPILOT: Okay.
CAPTAIN: Up, up, up.
CABIN: [*"Whoop, whoop, pull up."*]

END OF TAPE

Flight 965 first struck trees on the east side of a mountain named El Deluvio at 8,900 feet. The aircraft continued over the ridge and crashed into the summit on the west side of the mountain. The flight crew had gone off course and descended into a mountainous area at night. The captain and the copilot, it was determined, had committed a series "of operational errors that led to the accident." None of the errors when taken alone was egregious; however, when added together, they became fatal. The crew was hurried after it chose to land on Runway 19, rather than Runway 01, and necessary steps were performed improperly or not at all. They lost what is known as "situational awareness." They overlooked the fact of the terrain they were flying over and, finally and tragically, into.

Four passengers survived the crash with serious injuries.

Miami, the Everglades

May 11, 1996

■

ValuJet Flight 592

ValuJet. In the parlance of the controllers, the airline was "Critter," because of the alligator painted on its tail. It was a cut-rate airline, but there was nothing cut-rate about the crew of Flight 592 that spring afternoon as it pushed back and taxied out for a flight from Miami International Airport to Atlanta, Georgia. The day was bright, with thundershowers to the north. The flight departed at 1:44 P.M., a little late because of bad weather in Atlanta.

We start the CVR just as the cabin crew is preparing the passengers for pushback and departure.

FLIGHT ATTENDANT: [*On public-address system*] . . . The cabin has been pressurized for your comfort. If oxygen is [needed], reach up, pull down on the mask until the tubing is fully extended. Place the mask over your nose and mouth, secure it with the elastic band, and breathe normally. The oxygen bag may not appear to inflate. However, oxygen is flowing. For those of you traveling with small children, adjust your mask first and then assist the child. And passenger-seat cushions on this aircraft may be used as a flotation device and detailed instructions may be found on the safety information card. Smoking is not permitted at any time while on board this aircraft. . . . Also keep in mind, due to possible interference with navigational or communications systems, the following electronic devices may not be used during takeoff or landing: portable compact-disc players, portable computers, and cellular phones, which should be in the off position and stowed.

Now in preparation for takeoff, please fasten your seat belt, return your seat back and tray tables to the full upright and locked position. Your carry-on luggage must be stowed in the overhead compartments or underneath the seat in front of you. On behalf of all ValuJet employees, we'd like to thank you for selecting us today. We hope you enjoy your flight.

COPILOT: [*Coughs*] Shoulder harness?
CAPTAIN: On.
COPILOT: Parking brakes.
CAPTAIN: Set.
COPILOT: Fuel quantity.
CAPTAIN: Twenty-two three [2,300 pounds].
COPILOT: Twenty-two three. Pneumatic cross feeds?
PILOT: Open.
COPILOT: Anticollision light.
CAPTAIN: On.
COPILOT: Air-conditioning supply switches?
CAPTAIN: Off.
COPILOT: Fuel-boost pumps?
CAPTAIN: On.
COPILOT: Ignition?
CAPTAIN: On.
COPILOT: Before-start checklist complete.
FLIGHT ATTENDANT: Do you want me to tell them [passengers] information about [flight] connections?
CAPTAIN: You can tell them as we get closer. . . .
COPILOT: Oil pressure and N one.
FLIGHT ATTENDANT: And, ladies and gentlemen, for those of you that are all concerned about your connections, Captain Kubeck just informed me as soon as we get a little bit closer to Atlanta, we're able to relay . . . them [to you]. . . . We'll give you more information as we get it. Thank you.
CAPTAIN: . . . Okay, we'll be ready [to go] in about two minutes.
MIAMI GROUND CONTROL: Okay, thanks.
COPILOT: Uuuuuh.

The crew starts the engines.

CAPTAIN: Single-engine taxi, please.
COPILOT: Single engine . . . APU electric and air is set. Air-

conditioning, uh, supply on the right side is HP off. Air-conditioning auto-shutoff is [in] override. Ice-protection number two is off. Right pneumatic cross feed open. Number-two fuel-control level is off. [*Coughs*]

CAPTAIN: Before-takeoff check . . .

COPILOT: Performance data and bugs, for a hundred and five thousand, flaps five, Vee two [is at] one forty-six [mph].

CAPTAIN: One forty-six set and . . .

COPILOT: Trim?

CAPTAIN: . . . And, uh, one point nine nine across.

COPILOT: Yeah.

The crew goes through its complete set of pre-takeoff checks.

CAPTAIN: Okay, um, let's see. Where do we go? Three twenty-two off Dolphin [Miami]. It's a vector to Dolphin. . . . And I'll set three twenty-two.

COPILOT: [*Coughs*]

The crew discusses headings and weather.

CAPTAIN: Okay . . . it's my takeoff, it's my abort, flaps are five, Vee one [takeoff] min[imum] is one thirty-six [mph]. Runway heading to five. If we lose [power] on takeoff, depending which runway we go on, we'll, uh, do a visual back down the runway. . . . When we're about number three for takeoff, we'll start number two [engine to save on fuel]. Questions? No question?

COPILOT: No questions.

CAPTAIN: Is that an old Connie [Constellation] back there behind those buildings?

COPILOT: Ah, actually, it's one of those L-Ten-elevens.

CAPTAIN: Just behind this dirt pile here.

COPILOT: Okay, yeah.

CAPTAIN: [It] doesn't even have paint on it. Twin tails right there. About two o'clock. Hold short of Zulu. See it right there, off to the right?

COPILOT: Ah, I see a [Boeing] Seven forty-seven there.

CAPTAIN: He's real close to us, right behind us. The antenna's off to our right here by that building . . . it's kinda camouflaged.

COPILOT: Oh, this, this twin-tailed thing here?

CAPTAIN: Yeah.

COPILOT: Oh, that's a, uh, that's, uh . . .
CAPTAIN: A Connie [Constellation]?
COPILOT: A Connie One twenty-one.
CAPTAIN: Yeah.
COPILOT: Son of a gun.
CAPTAIN: You want to test out the PA and talk to the people? Um, you can ask if they hear or whatever you want. Understood whatever you use for words.
COPILOT: [*Coughs*] Ladies and gentlemen, from the cockpit, uh, we're on a hold right now for crossing traffic on the ground. Ah, and we only anticipate maybe a five- or ten-minute ground delay here, but, uh, we are cleared to Atlanta just as soon as they can get us out through all the traffic here in Miami.
FLIGHT ATTENDANT: [*On the interphone from the cabin*] There's a woman [standing] up. She wants to use the bathroom.
CAPTAIN: Yeah, [but tell her to] hurry.
COPILOT: Did you say warm it up?
CAPTAIN: No, a woman's using the restroom. I said hurry it up.
COPILOT: Aah.
CAPTAIN: Engine-restart checklist.
COPILOT: [*Coughs*] Left pneumatics closed, right pneumatic is open, right air-conditioning supply switch off. APU air is on. Fuel pumps are on. Ignition's on. Before-start checklist complete.
CAPTAIN: Ignition is on.
COPILOT: Number two [engine].
CAPTAIN: [*On interphone to flight attendant*] Is she [the woman who was going to the bathroom] still up?
FLIGHT ATTENDANT: Yes, she's up.
CAPTAIN: Nobody else [gets] up from here on in.
COPILOT: Oil pressure.
CAPTAIN: Check.
CABIN: [*Sound of increasing rpm*]
COPILOT: Lights out.

The crew goes through the restart checklist for engine number two.

CAPTAIN: If we have to go back, we'll back it up with the ILS. If we have to abort, you'll tell the Tower we're aborting.
COPILOT: Okay.
CAPTAIN: We're next up [for takeoff].

COPILOT: Yeah.
COPILOT: That's gotta be frustrating for those American [Airlines] guys. [They] have to wait for the company to give them their takeoff data.
CAPTAIN: I'd kinda like to have that problem. [*Chuckles*] Position and hold below the line.
COPILOT: [*To passengers*] Ladies and gentlemen, we've been cleared to the runway for departure. Flight attendants, please be seated. [*To captain*]: Antiskid coming on?
CAPTAIN: Yep.
CABIN: [*Sound of cockpit door being closed*]
COPILOT: Cockpit door is locked, ignition on, APU is off, flight-attendant signal is given, antiskid is armed, and TA, transponder is TA RA, pneumatic cross feeds are closed. . . .
CAPTAIN: Okay.
COPILOT: . . . Annunciator panel is checked, takeoff briefing is complete. Flood and logo lights.
CAPTAIN: Okay. Lights are on.
CABIN: [*Sound of engines increasing in rpm*]
CAPTAIN: Bleeds are closed. Set takeoff power.
COPILOT: Power is set. We have ninety-five [mph], ninety-four [knots of airspeed].
CABIN: [*Sound of aircraft nose tire riding over bumps in runway*]
COPILOT: [A] hundred knots.
CAPTAIN: Check.
COPILOT: Vee one.
CAPTAIN: Check.
COPILOT: Vee R [rotate]. Positive rate.
CAPTAIN: Gear up.
COPILOT: Vee two. Five [on flaps].
CAPTAIN: Flaps up.
COPILOT: [*To Tower*] Afternoon, Departure, Critter Five ninety-two is out of five hundred [feet] going to five thousand [feet].
MIAMI TOWER: Critter Five ninety-two, Departure. Good afternoon. Radar contact, climb and maintain seven thousand [feet].
COCKPIT: Seven thousand, Five nine two.
CAPTAIN: Slats retract.
COPILOT: Slats retract. Following the Boeing straight ahead of us in a turn.

CAPTAIN: Okay.

FLIGHT ATTENDANT: [*On public-address system*] And, ladies and gentlemen, Captain Kubeck will turn off the FASTEN SEAT BELT sign just as soon [as] she feels it's safe for you to get up and move about the cabin. Until that time, please remain comfortably seated with your seat belts securely fastened. We also suggest for your safety that seat belts be fastened even after the sign has been turned off. Shortly we will begin our in-flight service. We are pleased to offer you a variety of soft drinks, coffee, and juices. Cocktails are available for three dollars. Beer and wine are available for two dollars. As always, correct change is greatly, greatly appreciated. For a complete listing of our complimentary beverages, they may be found on page five of our *Good Times* magazine. For now, we just ask that you sit back and relax and enjoy your flight to Atlanta, Georgia.

TOWER: Critter Five ninety-two, turn left heading three six zero.

COPILOT: Three six zero, Five ninety-two.

CAPTAIN: Three six zero, climb power, climb check.

COPILOT: Power's set. Gear's up and checked, flaps up, lights out, spoilers [are] disarmed, the ignition is off, fuel pumps are set, air-conditioning shutoff is [in] override, hydraulic pumps off and low, flood and logo lights at ten, and altimeters at eighteen.

CAPTAIN: Thank you.

TOWER: Critter Five ninety-two, turn left heading three three zero.

CAPTAIN: Three three zero, Five nine two. Six [thousand feet] for seven.

COPILOT: Six for seven.

CABIN: [*Sound of altitude-alert signal*]

TOWER: Critter Five ninety-two, turn left heading three zero zero [and] join the Winco transition climb and maintain one six thousand [16,000 feet].

COPILOT: [*Coughs*]

CAPTAIN: Okay, let's turn on the radar. . . .

COPILOT: [We've] got something [a weather condition] out there about eighty miles out.

CAPTAIN: Okay.

COPILOT: That must be that thunderstorm.

CAPTAIN: There's a break here like . . . I'd hate to be in this thing at a hundred and eight thousand and in through this weather.

COPILOT: Yeah.

COPILOT: You don't want to hold them [the passengers in their seat belts] for a while?

CAPTAIN: No, they're okay for right now.

COPILOT: Flight attendants, departure check, please.

CABIN: [*Sound of beep on public-address/interphone channel*]

CAPTAIN: What was that?

COPILOT: I don't know.

CAPTAIN: We got some electrical problem.

COPILOT: Yeah. That battery charger's kickin' in. Ooh, we gotta . . .

CAPTAIN: We're losing everything.

TOWER: Critter Five nine two, contact Miami Center on one thirty-two forty-five, so long.

CAPTAIN: We need, we need to go back to Miami.

CABIN: [*Through the cockpit door, sounds of shouting from passenger cabin*]

FEMALE VOICE IN CABIN: [*Yelling through the door*] Fire, fire, fire, fire!

MALE VOICE IN CABIN: [*Yelling through the door*] We're on fire, we're on fire!

CABIN: [*Sound of landing-gear warning horn for three seconds*]

TOWER: Critter Five ninety-two, contact Miami Center, one thirty-two forty-five.

CAPTAIN: [We are returning] to Miami.

COPILOT: Uh, Five ninety-two needs immediate return to Miami.

TOWER: Critter Five ninety-two, uh, roger, turn left heading two seven zero. Descend and maintain seven thousand.

CABIN: [*Sounds of shouting from passenger cabin subside*]

COPILOT: Two seven zero, seven thousand, Five ninety-two.

TOWER: What kind of problem are you havin'?

CAPTAIN: Fire!

COPILOT: Uh, smoke in the cockp[it] . . . smoke in the cabin.

TOWER: Roger.

CAPTAIN: What altitude?

COPILOT: Seven thousand [feet].

CABIN: [*Sound of cockpit door opening; sound of six chimes, like cabin-service interphone*]

FLIGHT ATTENDANT: [*To cockpit*] Okay, we need oxygen. We can't get oxygen back here. Is there a way we could test them? [*Clears her throat*]

TOWER: Critter Five ninety-two, when able to turn, left heading two five zero. Descend and maintain five thousand.

CABIN: [*Sound of chimes, like cabin-service interphone; sounds of shouting from passenger cabin*]

COPILOT: Two five zero seven thousand.

FLIGHT ATTENDANT: [We're] completely on fire.

CABIN: [*Sounds of shouting, screaming from passenger cabin subside*]

COPILOT: [We are] outta nine [thousand feet].

CABIN: [*Sound of loud rushing air*]

COPILOT: Critter Five ninety-two, we need the, uh, closest airport available. . . .

TOWER: Critter Five ninety-two, they're [the emergency vehicles are] going to be standing by for you.

There is now an interruption of one minute and twelve seconds in CVR recording.

COCKPIT: [We] need radar vectors.

TOWER: Critter Five ninety-two, turn left heading one four zero.

COCKPIT: One four zero.

CABIN: [*Sound of loud rushing air*]

TOWER: Critter Five ninety-two, contact Miami Approach on corrections . . . no you . . . you just keep my frequency.

CVR: [*Repeating tones of recorder self-test signal start and continue; sound of rushing air*]

TOWER: Critter Five ninety-two, you can, uh, turn left heading one zero zero and join the Runway One two localizer at Miami.

END OF TAPE

ValuJet Flight 592 crashed in the Everglades swamp about eighteen miles northwest of the Miami Airport, from which it had departed minutes before. A fire had erupted in the forward cargo hold in oxygen generators, classified as HAZMAT (Hazardous Material) and used in aircraft passenger-service units. The heat of the fire was intense, estimated at nearly 3,000 degrees F., and spread through the entire cabin.

Everyone on board—105 passengers, 2 flight attendants, a crew of

2—died either on impact or from fire-related injuries. The airplane plummeted from 10,000 feet nose down, reportedly because either the captain or the copilot was slumped against the control yoke.

What made the wreckage particularly grim: A swamp swallowed up the ValuJet DC-9 without a trace. The aircraft could not be found immediately because the impact of the hull of the Douglas DC-9-32 against the ground left no scar on the terrain.

Pittsburgh, Pennsylvania

September 8, 1994

■

USAir Flight 427

The crash of USAir Flight 427, a Boeing 737-300, stumped crash investigators for nearly three years before they were certain enough of their conclusions to make them a matter of public record. For many months after the crash, what happened in the skies near Pittsburgh remained a total mystery. An airplane carrying 127 passengers and a crew of 5 (2 pilots) simply and without warning rolled over and dove straight into the ground while on its initial approach to land.

The day was clear; the flight was on a visual approach. The time was around seven in the evening. The nearest other aircraft was four and a half miles away with 1,500 feet of vertical separation. The crew was in a jovial, even festive, mood. The captain yawned as the tape began. Landing that evening was a little bit like bringing a train into a station.

We begin the CVR about nineteen minutes from the flight's meeting with disaster.

FLIGHT ATTENDANT: [*To the cockpit crew*] They didn't give us connecting flight information or anything. Do you know what gate we're coming in to?
CAPTAIN: Not yet.
FLIGHT ATTENDANT: Any idea?
CAPTAIN: No.
FLIGHT ATTENDANT: Do ya know what I'm thinkin' about? Pretzels.
CAPTAIN: Pretzels?

FLIGHT ATTENDANT: [Do] you guys need drinks here?
CAPTAIN: I could use a glass of somethin', whatever's open, water, uh, water—a juice?
COPILOT: I'll split a, yeah, a water, a juice, whatever's back there. I'll split one with him.
FLIGHT ATTENDANT: Okeydokey. Do you want me to make you my special fruity juice cocktail?
CAPTAIN: How fruity is it?
FLIGHT ATTENDANT: Why don't you just try it.
COPILOT: All right, I'll be a guinea pig.
CABIN: [*Sound of cabin door closing as flight attendant leaves the cockpit to get the drinks*]

Flight 427 is instructed to reduce its speed to 210 knots and maintain 10,000 feet altitude. A few seconds later, the flight is instructed to contact Pittsburgh Approach on frequency 121.25.

CAPTAIN: Two ten, he said?
COPILOT: Two ten? Oh, I heard two fifty. . . .
CAPTAIN: I may have misunderstood him.

Pittsburgh Approach asks them to turn left, heading 100.

CABIN: [*Sound of cockpit door opening*]
FLIGHT ATTENDANT: Here it is [the special fruity cocktail mixture].
CAPTAIN: All right.
COPILOT: All right. Thank you. Thank you.
FLIGHT ATTENDANT: . . . I didn't taste 'em, so I don't know if they came out right.
CAPTAIN: That's good.
COPILOT: That *is* good.
FLIGHT ATTENDANT: It's good.
COPILOT: That is different. [It'd] be real good with some dark rum in it.
FLIGHT ATTENDANT: Yeah, right.
PITTSBURGH APPROACH: USAir Four twenty-seven, Pittsburgh Approach. Heading one six zero, vector ILS Runway Twenty-eight right final approach course. Speed two one zero.
COPILOT: What kind of speed?
CAPTAIN: We're comin' back to two ten [210 knots] and, uh, one sixty heading, down to ten [10,000 feet], USAir Four twenty-

seven. . . . [*To copilot*] What runway did he say? [*To flight attendant*] It tastes like a . . .

COPILOT: Good.

CAPTAIN: There's [a] little grapefruit in it?

FLIGHT ATTENDANT: No.

COPILOT: Cranberry?

FLIGHT ATTENDANT: Yeah. You saw that from the color.

CAPTAIN: [What] else is in it?

COPILOT: Uh, Sprite?

FLIGHT ATTENDANT: Diet Sprite.

COPILOT: Huh.

FLIGHT ATTENDANT: And I guess you could do with Sprite. Probably be a little better if you do.

CAPTAIN: Yeah. There's more?

FLIGHT ATTENDANT: One more.

COPILOT: Ah. OJ?

FLIGHT ATTENDANT: You got it.

COPILOT: Huh?

FLIGHT ATTENDANT: Cranberry, orange, and Diet Sprite.

COPILOT: Really nice.

FLIGHT ATTENDANT: It's different. . . .

CAPTAIN: I always mix cranberry and the grapefruit. I like that.

FLIGHT ATTENDANT: Okay, back to work.

COPILOT: Okay.

CABIN: [*Sound of cockpit door opening and closing*]

COPILOT: I suspect we're going to get the right side.

APPROACH: USAir Four twenty-seven, descend and maintain six thousand.

CAPTAIN: Cleared to six, USAir Four twenty-seven.

COPILOT: Oh, my wife would like that [drink].

CAPTAIN: Cranberry, orange, and Sprite.

COPILOT: Yeah. I guess we ought to do a preliminary.

The crew now goes through their pre-landing checks. Approach requests a left turn heading 140, and to reduce their speed to 190 knots.

CABIN: [*Sound of flap handle being moved; sound of seat belt chime*]

COPILOT: Oops. I didn't kiss 'em [the passengers] good-bye. What was the temperature? [Do you] remember?

CAPTAIN: Seventy-five.

COPILOT: Seventy-five?

FLIGHT ATTENDANT: [*On public-address system*] . . . Seat belts and remain seated for the duration of the flight.

COPILOT: [*On public-address system*] Folks, from the flight deck, we should be on the ground in about ten more minutes. Uh, sunny skies, a little hazy. Temperature . . . temperature's, ah, seventy-five degrees. Wind's out of the west around ten miles per hour. [We] certainly 'ppreciate you choosing USAir for your travel needs this evening. Hope you enjoyed the flight. Hope you come back and travel with us again. This time we'd like to ask our flight attendants [to] please prepare the cabin for arrival. Ask you to check the security of your seat belts. Thank you.

CABIN: [*Seat-belt chime*]

CAPTAIN: Did you say [Runway] Twenty-eight left for USAir Four twenty-seven?

APPROACH: Uh, USAir Four twenty-seven, it'll be Twenty-eight right.

CAPTAIN: Twenty-eight right, thank you. [*To copilot*] Twenty-eight right.

COPILOT: Right, Twenty-eight right. That's what we planned on. Auto-brakes on one for it.

Throughout, Pittsburgh Approach is issuing instructions to other aircraft in the area.

CAPTAIN: Seven [thousand] for six [thousand feet].

COPILOT: Seven for six.

CAPTAIN: Boy, they [Pittsburgh Approach] always slow you up so bad here.

COPILOT: That sun is gonna be just like it was takin' off in Cleveland yesterday, too. I'm just gonna close my eyes. [*Laughs*] You holler when it looks like we're close. [*Laughs*]

CAPTAIN: Okay.

APPROACH: USAir Four twenty-seven, turn left heading one zero zero. Traffic will be one to two o'clock, six miles [away], [and it will be a] northbound Jetstream [which is] climbing out of thirty-three [3,300 feet] for five thousand [feet].

CAPTAIN: We're looking for the traffic, turning to one zero zero, USAir Four twenty-seven.

CABIN: [*Sound of engines increasing rpms*]
COPILOT: Oh, yeah. I see the Jetstream.
CAPTAIN: Sheez . . .
COPILOT: Huh?
CABIN: [*Sound of thump; sound like "clickety-click"; again the thumping sound, but quieter than before*]
CAPTAIN: Whoa . . . Hang on.
CABIN: [*Sound of engines increasing rpm; "clickety-click" sound; sound of trim wheel turning at autopilot trim speed; sound of pilot grunting; sound of wailing horn similar to autopilot-disconnect warning*]
CAPTAIN: Hang on.

Flight 427 is now turning over on its back at 6,000 feet and heading for the ground at 300 miles an hour. It would take another sixteen seconds before the aircraft hit the ground.

COPILOT: Oh, shit.
CAPTAIN: Hang on. What the hell is this?
CABIN: [*Sound of stick shaker; sound of altitude alert*]
CABIN: "Traffic. Traffic." [*Sound through the cabin door*]
CAPTAIN: What the . . .
COPILOT: Oh . . .
CAPTAIN: Oh, god, oh, god . . .
APPROACH: USAir . . .
CAPTAIN: Four twenty-seven, emergency!
COPILOT: [*Screams*]
CAPTAIN: Pull . . .
COPILOT: Oh . . .
CAPTAIN: Pull . . . Pull . . .
COPILOT: God . . .
CAPTAIN: [*Screams*]
COPILOT: No . . .

END OF TAPE

The airplane struck the ground near Aliquippa, Pennsylvania, in a slight roll to the left at an eighty-degree angle of descent, nearly perpendicular, going an estimated 299 mph. At the point of impact, the wreckage was buried eight feet deep in the earth. The aircraft hull disintegrated, and there was an intense fire. Also at the time of impact, both engines were running symmetrically and powered with thrust. The thrust reversers were stowed. The flaps were at a "Flap 1" setting, and spoilers and landing gear were retracted. These were all expected positions for an aircraft during its initial approach for landing.

All on board died instantly.

The mystery of Flight 427 led to the longest and most difficult crash investigation in the history of the NTSB, which has not determined the cause of the accident as of this writing, though it suspects that the trail of this tragic event was started by the wake vortex of another aircraft in the seconds before the accident began. That was followed by a large rudder movement in Flight 427, perhaps to compensate for the wake vortex, which then spun the aircraft out of control.

Kahului, Hawaii

April 28, 1988

■

Aloha Airlines Flight 243

On a regularly scheduled flight from Hilo to Honolulu, Hawaii, Aloha Airlines Flight 243 took off and climbed to a cruise altitude of 24,000 feet. It was at that flight level when the ceiling area of the forward passenger cabin suddenly burst open in an explosive decompression. The ceiling separated from the airplane, leaving the passengers from the cockpit door to the front of the wing exposed to the elements as if they were in a convertible car. Riding along in terror, they could do nothing as the aircraft dove to an altitude level (around 11,000 feet) where oxygen was not needed. There were two pilots aboard that early afternoon, an observer in the cockpit jumpseat, three flight attendants, and eighty-nine passengers.

We pick up the CVR just as the ceiling rips off.

CABIN: [*Sound of screams; sound of wind noise*]

The CVR microphones in the cockpit could not pick up any crew conversation for the next five minutes. However, the CVR recorded the crew's transmissions with Ground Control through the crew's oxygen-mask microphones.

COPILOT: Center, Aloha Two forty-three. We're going down. . . . Request lower [altitude]. Center, Aloha Two forty-three. Center, Aloha Two forty-three. Maui Approach, Aloha Two forty-three. Maui Tower, Aloha Two forty-three. Maui Tower, Aloha Two forty-three. We're inbound for a landing. Maui Tower, Aloha Two forty-three.

MAUI TOWER: [Flight] callin' Tower, say again.

COPILOT: Maui Tower, Aloha Two forty-three, we're inbound for landing. We're just, ah, west of Makena . . . just to the east of Makena, descending out of thirteen [13,000 feet], and we have rapid depr— We are unpressurized. Declaring an emergency. . . .

TOWER: Aloha Two forty-three, winds zero four zero at one five. Altimeter two niner niner niner. Just to verify again. You're breaking up. Your call sign is Two forty-four? Is that correct? Or Two forty-three?

Here the crew, having reached 11,000 feet, takes off its oxygen masks.

COPILOT: Two forty-three Aloha—forty-three.

TOWER: Two forty-two, the equipment is on the roll. Plan [to approach] straight in [on] Runway Two, and I'll keep you advised on any wind change.

COPILOT: Aloha Two forty-three . . . [*To captain*] Do you want me to call for anything else . . . ?

CAPTAIN: Nope.

TOWER: Is that Aloha Two forty-four on the emergency?

COPILOT: Aloha Two forty-three.

TOWER: Ah. Two forty-six.

COPILOT: Aloha Two forty-three.

TOWER: Aloha Two forty-three, say your position.

COPILOT: We're just, ah, to the east of Makena [at a] point descending out of eleven thousand [11,000 feet]. Request clearance into Maui for landing. Request the [emergency] equipment.

TOWER: Okay, the equipment is on the field . . . is on the way. Squawk zero three four three. Aloha Two forty-three, can you come up on [frequency] one one niner point five?

COPILOT: Two forty-three. Can you hear us on one nineteen five two? Maui Tower, Two forty-three. It looks like we've lost a door. We have a hole in this, ah, left side of the aircraft.

JUMPSEAT PASSENGER: I'm fine.

COPILOT: [*To captain*] Want the [landing] gear?

CAPTAIN: No.

COPILOT: Want the [landing] gear?
CAPTAIN: No.
COPILOT: Do you want it [the gear] down?
CAPTAIN: Flaps fifteen [for] landing.
COPILOT: Okay.
CAPTAIN: Here we go. We've picked up some of your airplane business right there. I think that they can hear you. They can't hear me. Ah, tell him, ah, we'll need assistance to evacuate this airplane.
COPILOT: Right.
CAPTAIN: We really can't communicate with the flight attendants, but we'll need trucks, and we'll need, ah, air stairs from Aloha.
COPILOT: All right. [*To Tower*] Maui Tower, Two forty-three, can you hear me on Tower?
TOWER: Aloha Two forty-three, I hear you loud and clear. Go ahead.
COPILOT: Ah, we're gonna need assistance. We cannot communicate with the flight attendants. Ah, we'll need assistance for the passengers when we land. . . .
TOWER: Okay, I understand you're gonna need an ambulance. Is that correct?
COPILOT: Affirmative.
CAPTAIN: [*To copilot*] It feels like manual reversion.
COPILOT: What?
CAPTAIN: Flight controls feel like manual reversion [like the autopilot has switched off].
COPILOT: Can we maintain altitude okay?
CAPTAIN: Ah . . .
COPILOT: Can we maintain altitude okay?
CAPTAIN: Let's try flying . . . let's try flying with the gear down here.
COPILOT: All right, you got it.
CABIN: [*Sound of gear being lowered*]
TOWER: Aloha Two forty-three, can you give me your souls on board and your fuel on board?
CAPTAIN: [*To copilot*] Do you have a passenger count for the Tower?
COPILOT: [*To Tower*] We, ah—eighty-five, eighty-six, plus five crew members.
TOWER: Okay. And, ah, just to verify. You broke up initially. You do need an ambulance. Is that correct?
COPILOT: Affirmative.

TOWER: Roger. How many do you think are injured?
COPILOT: We have no idea. We cannot communicate with our flight attendant.
TOWER: Okay. We'll have an ambulance on the way.
COPILOT: There's a possibility that, ah, we won't have a nose gear.
CAPTAIN: [*To copilot*] You tell 'em that we got such problems, but we are going to land anyway—even without the nose gear. But they should be aware of . . . we don't have a safe nose-gear-down indication.
TOWER: Aloha Two forty-three, wind zero five. The [emergency] equipment is in place.
COPILOT: Okay, be advised. We have no nose gear. We are landing without nose gear.
TOWER: Okay. If you need any other assistance, advise. . . .
COPILOT: We'll need all the [emergency] equipment you've got. [*To captain*] Is it easier to control with the flaps up?
CAPTAIN: Yeah. Put 'em at five. Can you give me a Vee speed for a flaps-five landing?

The copilot can't find the Vee speed.

COPILOT: Do you want the flaps down as we land?
CAPTAIN: Yeah, after we touch down.
COPILOT: Okay.

The captain and copilot discuss the speeds for landing.

TOWER: Aloha Two forty-three, just for your information. The gear appears down. Gear appears down.
COPILOT: [*To captain*] Want me to go flaps forty . . . ?
CAPTAIN: No.
COPILOT: Okay.
CABIN: [*Sound of touchdown on runway*]
COPILOT: Thrust reverser.
CAPTAIN: Okay. Okay, shut it down.
COPILOT: Shut it down.
CAPTAIN: Now, left engine.
COPILOT: Flaps.
TOWER: Aloha Two forty-three, just shut her down where you are. Everything [is] fine. The gear did. . . . The fire trucks are on the way. . . .

Captain: Okay.
Cabin: [*Sound of engines winding down*]
Captain: Okay, start the call for the emergency evacuation.

END OF TAPE

The Boeing 737 of Aloha Flight 243 was manufactured in 1969 and had accumulated 35,496 flying hours and 89,680 takeoff-landing cycles. The cause of the separation of the aircraft's ceiling was attributed to static overstress. The airplane was old, and the cycles of pressurization and depressurization had weakened parts of the fuselage. One flight attendant was killed. All the passengers landed safely, though sixty-five of them were injured to varying degrees, mostly minor.

Sioux City, Iowa

July 19, 1989

■

United Airlines Flight 232

I have chosen to end the CVR collection with this tape because of what it says about the professionalism and sometimes heroism of airlines' crews.

United Airlines Flight 232, a DC-10, was on a regularly scheduled flight from Denver to Chicago carrying 296 passengers, 8 flight attendants, and 3 crewmen in the cockpit, including Captain Al Haynes, a fifty-eight-year-old veteran of commercial aviation.

While flying at a flight level of 37,000 feet at 3:16 P.M. about two hundred miles west of Dubuque, Iowa, Flight 232 experienced an explosive failure of its number-two, or tail, engine. The separation of the engine, the fragmentation of its fan blades and other parts, and the explosive force of the fan rotor parts severed the aircraft's hydraulic lines, which controlled the wing flaps, elevators, ailerons, and rudder. In other words, all steering mechanisms were lost. For the next forty-one minutes, Haynes and his crew maneuvered the airplane with the throttles of the two remaining wing engines. Haynes turned the aircraft in large arcing spirals from 33,000 feet to land at Sioux City, Iowa. The most amazing thing about him, as the CVR demonstrates, is that he did not lose his sense of humor in the face of the emergency. He knew the predicament, yet he rose above it, especially when a representative of United Airlines' System Aircraft Maintenance in San Francisco refused to believe what the crew of Flight 232 was telling him; his insistence that all three hydraulic systems could not have been destroyed became a running joke for the flight crew in the ongoing emergency.

CAPTAIN HAYNES: [*To Sioux City Approach*] Ah, we're controlling the turns by power. I don't think we can turn right. I think we can only make left turns. We're starting a little bit of a left turn right now. Maybe we can only turn right. We can't turn left.

SIOUX CITY APPROACH: United Two thirty-two heavy, ah, [I] understand you can only make right turns.

HAYNES: That's affirmative.

APPROACH: United Two thirty-two heavy, roger. Your present track puts you about eight miles north of the airport, sir. And, ah, the only way we can get you around [Runway 31] is a slight left turn with differential power or if you go and jockey it over.

HAYNES: Roger. Okay, we're in a right turn now. It's about the only way we can go. We'll be able to make very slight turns on final, but right now just . . . we're gonna make right turns to whatever heading you want.

APPROACH: United Two thirty-two heavy, roger. Ah, right turn, heading two five five.

HAYNES: Two five. [*To copilot*] Now the goddamn elevator doesn't want to work. Rolling right.

FLIGHT ENGINEER: [*On radio to United Airlines System Aircraft Maintenance (SAM)*] This is United Two thirty-two. We blew number-two engine, and we've lost all hydraulics and we are only able to control, ah, level flight with the, ah, asymmetrical power settings. We have very little rudder or elevator.

COPILOT: [*To flight engineer*] Very little elevator. [It's] hard or sluggish. [*To Haynes*] Ah, Al, do you want me to slew this elevator? Feel [it].

HAYNES: Yeah, whatever you can [do].

APPROACH: United Two thirty-two heavy, fly heading two four zero and say your souls on board.

COPILOT: [*To Haynes*] Ah, now the nose is coming up.

SAM: United Two thirty-two, understand that you lost number-two engine totally, sir?

HAYNES: Say again.

APPROACH: Souls on board, United Two thirty-two heavy.

HAYNES: Gettin' that right now.

FLIGHT ENGINEER: [*To SAM*] That's affirmative.

SAM: Your, ah, system one and system three—are they operating normally?

FLIGHT ENGINEER: Negative. All hydraulics are lost. All hydraulic systems are lost. The only thing we have is the . . . [and he itemizes the systems that are working].

APPROACH: United Two thirty-two heavy, can you continue your turn to heading two four zero?

COPILOT: I don't know. We'll try for it.

SAM: Okay, United Two thirty-two, understand you have normal power on one and three engines.

FLIGHT ENGINEER: That's affirmative.

COPILOT: [I] wonder about the outboard ailerons. If we put some flaps out, you think that would give us outboard?

FLIGHT ENGINEER: God, I hate to do anything.

HAYNES: Well, we're going to have to do something.

SAM: United Two thirty-two, is all hydraulic quantity gone?

There is a growing sense of incredulity here on the part of the SAM agent; losing all three hydraulic systems just isn't in the odds book, and he doubts the truth of what he is hearing.

FLIGHT ENGINEER: Yes, all hydraulic quantity is gone.

COPILOT: Level off.

APPROACH: United Two thirty-two heavy, souls on board and fuel remaining?

COPILOT: Souls on board and fuel remaining. We have thirty-seven six [on fuel].

FLIGHT ENGINEER: We've got thirty-seven four on fuel.

APPROACH: Roger.

CABIN: [*Sound of two knocks on cockpit door*]

SAM: Okay, United Two thirty-two, where you gonna set down?

HAYNES: Unlock the door.

The first flight attendant hears the engine explode while she is picking up trays and sits down on the floor until things seem normal again. After Captain Haynes makes an announcement to the passengers that the plane has lost an engine, she stands up and reassures a passenger sitting in seat 23A, who is clutching a baby, and a woman in 23E, who looks very frightened. Captain Haynes then calls the first flight attendant on the interphone and asks her to come to the cockpit. There he tells her the situation. She should prepare the cabin for evacuation. She does not ask any questions because, as she later said, "The cockpit crew was working very hard." She does not gather the flight attendants together because she does not want to alarm the passengers, and so she informs each of them individually. She tells them to secure everything as quickly as possible. The trays are picked up.

The first flight attendant is aware of a United pilot who is traveling in the first-class cabin in civilian clothes. A passenger points out the window to show her the right horizontal stabilizer, which is visibly damaged. She returns to the cockpit and describes what she has seen to Captain Haynes, who tells her that the landing will be "quick and dirty."

Her hands are shaking as she puts away meal trays, postponing the announcement of plans for the evacuation until she is "sure that I could stand there and do it in the manner that I am supposed to." She reads the quick-preparation speech to the passengers over the public-address system. She adds that infants on the airplane should be placed on the floor. She briefs passengers in seats 9A, 9B, and 9C on the operation of the exit door and asks two men to assist passengers at the bottom of the slide during the evacuation.

Meanwhile, back in the cockpit, the crew is struggling to get Flight 232 down safely.

HAYNES: What's SAM saying?

CABIN: [*Sound of three knocks on door*]

FLIGHT ENGINEER: We need some assistance right now. We can't . . . we're havin' a hard time controllin' it.

SAM: Okay, United Two thirty-two.

HAYNES: We don't have any controls.

COPILOT: You want to go forward on it, Al.

CABIN: [*Sound of two knocks on door; sound of landing-gear warning horn*]

HAYNES: Now go forward. . . . Now let it come back. Got to lead . . . got to lead it. . . .

SAM: I'll try to help ya. I'll pull out your flight manual.

Here Captain Haynes turns to address a captain riding in the cockpit jumpseat (referred to as Jumpseat Captain).

HAYNES: See what you can see back there [in the cabin], will ya?
COPILOT: Go back and look out [at] the wing . . . and see what we got. . . .
JUMPSEAT CAPTAIN: Okay.
HAYNES: [Pull] back on the sucker.
CABIN: [*Cockpit door opens*]
COPILOT: Don't pull the throttles off. . . .

For the next few seconds the copilot and Captain Haynes continue to maneuver with the throttles, while the flight engineer discusses the problem with SAM by radio.

COPILOT: What's the hydraulic quantity?
FLIGHT ENGINEER: Down to zero.
COPILOT: On all of them?
FLIGHT ENGINEER: [On] all of them.
HAYNES: Quantity, quantity is gone?
FLIGHT ENGINEER: Yeah, all the quantity is gone. All pressure is [gone].
HAYNES: [Did] you get ahold of SAM?
FLIGHT ENGINEER: Yeah, I've talked to him.
HAYNES: What's he saying?
FLIGHT ENGINEER: He's not telling me anything.
HAYNES: We're not going to make the runway, fellas. We're gonna have to ditch this son of a bitch and hope for the best.
CABIN: [*Sound of three knocks*]
HAYNES: Unlock the damn door.
COPILOT: Unlock it.
HAYNES: We've lost . . . no hydraulics. We have no hydraulic fluid. That's part of our main problem.
UNITED AIRLINES CHICAGO HEADQUARTERS DISPATCH: United Two thirty-two, do you want to put that thing on the ground right now, or do you want to come to Chicago?
FLIGHT ENGINEER: Okay, we're, ah, we don't know what we'll be able to do. We don't think we're even gonna be able to get on the runway right now. We have no control hardly at all. . . .
JUMPSEAT CAPTAIN: [*Returning to cockpit*] Okay, both your inboard ailerons are sticking up. That's as far as I can tell. I don't know. . . .
HAYNES: Well, that's because we're steering. . . . We're turning maximum turn right now.

Jumpseat Captain: Tell me. Yell what you want and I'll help you.

Haynes: Right throttle. Close one, put two up. What we need is elevator control. And I don't know how to get it.

Jumpseat Captain: Okay, ah . . .

Flight Engineer: [*To Dispatch*] Roger, we need any help we can get from SAM, as far as what to do with this. We don't have anything. We don't [know] what to do. We're having a hard time controlling it. We're descending. We're down to seventeen thousand feet. We have . . . ah, hardly any control whatsoever.

Haynes: [*To jumpseat captain*] The only help you can get is the autopilot, and I tried that and it won't work.

Jumpseat Captain: It won't work. Okay . . .

Haynes: Go ahead and try it again. Pull back, pull back, pull back.

Dispatch: Okay, copy that, Two thirty-two. San Fran [is on the] line. Give 'em all the help you can. We'll get you expedited handling into Chicago [to] put you on the ground as soon as we can. . . .

Haynes: You want full aileron and full elevator. No, no, no, no, no, not yet. Wait a minute. Wait till it levels off. Now go.

Flight Engineer: [*To Dispatch*] Well, we can't make Chicago. We're gonna have to land somewhere out here, probably in a field.

Haynes: How're they doin' on the evacuation [preparations]?

Jumpseat Captain: They're puttin' things away, but they're not in any big hurry.

SAM: United Two thirty-two, we [understand that you] have to land [at] the nearest airport, the nearest airport. Ah, I'm tryin' to find out where you've lost all three hydraulic systems.

Haynes: Well, they better hurry. We're gonna have to ditch, I think.

Jumpseat Captain: Yeah.

Haynes: Okay.

Cabin: [*Sound of knock on door*]

Haynes: I don't think we're going to make the airport.

Copilot: No. We got no hydraulics at all.

Cabin: [*Sound of landing-gear warning horn*]

Jumpseat Captain: Get this thing down. We're in trouble. . . .

Flight Engineer: [*To SAM*] That is affirmative. We have lost all three hydraulic systems. We have no quantity and no pressure on any hydraulic system. . . .

Haynes: [*To Sioux City Approach*] Sir, we have no hydraulic fluid, which means we have no elevator control, almost none, and very little aileron control. I have serious doubts about making the airport. Have you got someplace near there, ah, that we might be able to ditch? Unless we get control of this airplane we're gonna put it down wherever it happens to be.

SAM: Ah, United Two thirty-two, you have lost all manual flight-control systems?

Flight Engineer: That's apparently true.

SAM: United Two thirty-two, ah, in the flight manual [on page] sixty-three . . .

Haynes: Gotta put some flaps and see if that'll help. . . .

Flight Engineer: [*To SAM*] I am on [page] sixty-three.

Copilot: You want them [the flaps] now?

Haynes: What the hell. Let's do it. We can't get any worse than we are. . . .

Copilot: Slats are out?

Jumpseat Captain: No, you don't have any slats.

Haynes: We don't have any hydraulics, so we're not going to get anything.

Approach: United Two thirty-two heavy, can you hold that present heading, sir?

Haynes: This is Sioux City, Iowa. That's where we're headed.

There is confusion about where they are and how far they are from Sioux City.

Haynes: [*To Approach*] Where's the airport now for [United] Two thirty[-two]? We're turning around in circles.

Jumpseat Captain: [*Probably to copilot*] You get on number one and ask them where the hell we are.

Haynes: [*To Approach*] Where's the airport to us now, as we come spinning down here?

Approach: United Two thirty-two heavy, Sioux City Airport is about twelve o'clock and three six [thirty-six] miles.

Haynes: Okay. We're tryin' to go straight. We're not havin' much luck.

Jumpseat Captain: All right, I got you on seven hundred on the squawk, so they can track ya [on radar].

Copilot: He's got us on radar.

Haynes: As soon as the nose starts up we have to push forward on the yoke.

Jumpseat Captain: We got nothing on number two, number two [engine]?

Haynes: No, no, we got it shut down.

SAM: United Two thirty-two, I'm gettin' contact with Flight Ops right now. Stand by, please.

Haynes: I want a heading of about three zero zero. We kinda got level flight back again.

Jumpseat Captain: Okay, if you got denser air, you should [get level flight back again]. Whatever you got, you got.

Haynes: A little better.

Jumpseat Captain: Okay, ah, let me see. . . .

Haynes: [*Laughs*] We didn't do this thing on my last [performance check in a simulator].

Cabin: [*Laughter*]

Copilot: No.

Haynes: [I] poured coffee all over . . . it's just coffee. We'll get this thing on the ground. Don't worry about it.

Copilot: It seems controllable, doesn't it, Al?

Jumpseat Captain: Yeah. The lower you get the more dense that air is [and] the better your shots. Okay?

Haynes: I'll tell ya what we need. We're puttin' this thing into Sioux City. Get me. . . . [*To Approach*] Sioux City, United Two thirty-two, could you give us, please, your ILS frequency, the heading, and the length of the runway?

Approach: United Two thirty-two heavy, affirmative. The localizer frequency is one zero nine point three and you're currently about thirty-five miles to the northeast. It'll take about two two two three five two four zero heading to join it.

SAM: United Two thirty-two, this is SAM.

Cabin: [*Sound of landing-gear warning horn*]

Flight Engineer: [*To SAM*] Sam, Two thirty-two. We're gonna try and put into Sioux City.

Dispatch: SAM, this is Dispatch. I haven't been able to copy Two thirty-two. We're hearing a rumor that he's on approach to Sioux City Airport. Last we heard he's at seventeen thousand feet and he may be too low for us to maintain contact with him. Go ahead. . . .

Approach: United Two thirty-two, understand you are gonna try to make it into Sioux City. There's no airport out that way that can accommodate you, sir.

HAYNES: Okay, we'll head for Sioux City. We got a little bit of control back now. How long [is] your runway?

FLIGHT ENGINEER: [*To SAM*] Two thirty-two is very busy right now. We're tryin' to go into Sioux City. We'll call you as soon as I can. . . .

APPROACH: Two thirty-two heavy, the airport, the runway is nine thousand feet long. . . .

SAM: He has no control. He's using that kind of sink rate, I believe.This is what he's doing. He's got his hands full for sure.

HAYNES: Okay, thank you. [*To jumpseat captain*] You're a little more . . . Let's see if you can make a left turn.

JUMPSEAT CAPTAIN: Left turn. All right. Your speed is what? I'm worried about [it]. I don't want to stall you.

DISPATCH: [*To SAM*] Okay, we just wanted [Flight Two thirty-two] to say where he was so we can get emergency equipment out. Thanks a lot, SAM.

FLIGHT ENGINEER: [*To Haynes*] You want a no flap–no slat [landing], right?

HAYNES: Yeah. Ah, start dumpin' [fuel], will ya? Just hit the quick [fuel] dump. Let's get the weight down as low as [we] can. . . .

FLIGHT ENGINEER: I didn't have time to think about that [dumping fuel].

HAYNES: Try not to lose any more [altitude] than we have to.

COPILOT: What? Altitude?

HAYNES: Yeah.

COPILOT: Okay. Go ahead and dump.

JUMPSEAT CAPTAIN: He's got his weight. He's got his weight. He's only got about a thousand pounds [of fuel] to go.

COPILOT: Okay.

JUMPSEAT CAPTAIN: This thing seems to want to go right more than it wants to go left, doesn't it?

APPROACH: United Two thirty-two, did you get the souls on board count?

HAYNES: [*To crew*] What did you have for a count for people? [*To Approach*] [Let me] tell you, right now we don't even have time to call the gals [the flight attendants in the passenger cabin for a passenger count]. . . .

FLIGHT ENGINEER: Ah, two ninety-two.

APPROACH: Roger.

HAYNES: Ease all the power back.

JUMPSEAT CAPTAIN: Okay, [the] nose is coming up.

SAM: All hydraulic systems are gone. . . .

APPROACH: Okay, thank you.

COPILOT: Yeah, we're goin' up.

HAYNES: Yeah, I know it. I'm pushin' with all I got.

JUMPSEAT CAPTAIN: [The] power's coming back. Power's coming back. . . .

HAYNES: As soon as it starts to come back . . . Okay, come back. . . .

JUMPSEAT CAPTAIN: Power's comin' back in. . . .

HAYNES: Bring [turn] it to the right with the right one. You got to go left. We just keep turnin' right. Still turnin' right . . .

JUMPSEAT CAPTAIN: That's what I'm tryin' to do. . . .

HAYNES: [*To Approach*] Two thirty-two, we're just gonna have to keep turnin' right. There's not much we can do about [turning] left. We'll try to come back around to the heading. . . .

COPILOT: Is this Sioux City down to the right?

HAYNES: That's Sioux City.

APPROACH: United Two thirty-two, roger. [We] need you on about a two three five heading, sir, if you can manage that and hold that.

HAYNES: Well, we'll see what happens. . . .

JUMPSEAT CAPTAIN: [We're] goin' down now. I'll put a little bit more power [on]. I'm gonna try and set about ninety percent [and] see if that holds up good for you. Tryin' to find the right power setting so you don't have to fight this pitch.

COPILOT: Ease it to the right.

HAYNES: Did you ever get hold of SAM?

FLIGHT ENGINEER: Yep. Didn't get any help.

HAYNES: [*Laughs*] Okay, did you tell 'em to advise Dispatch of our situation and what we're doing?

FLIGHT ENGINEER: Yes, he knows.

HAYNES: No more right turns, no more. Ah, I mean, ah, we want to turn right. He [Sioux City Approach] wants us to turn right.

JUMPSEAT CAPTAIN: You do want to turn . . . all right?

HAYNES: [*Exhales*]

COPILOT: [*To Approach*] Where is Sioux City from our present position, United Two thirty-two?

APPROACH: United Two thirty-two, it's about twenty [degrees] on the heading and thirty-seven miles. . . .

COPILOT: [There is an] airport right below us here, but . . .

Haynes: They said it won't accommodate us.
Copilot: Okay.
Haynes: See if you can keep us with the throttles in a ten- to fifteen-degree turn. . . .
Jumpseat Captain: All right. I'll play 'em [throttles]. I'll play 'em. I'll power up this number-three engine and try to accommodate you.
Haynes: You had the thing leveled off for a minute.

Denny Fitch, a training pilot for United who has been riding in first class, comes into the cockpit to see whether he can assist the crew. The jumpseat captain gives him his place, and he goes back into the cabin and takes a seat in the rear of the airplane.

Haynes: My name's Al Haynes.
Fitch: Hi, Al. Denny Fitch.
Haynes: How do you do, Denny?
Fitch: I'll tell you what. We'll have a beer when this is all done.
Haynes: Well, I don't drink, but I'll sure as hell have one. Little right turns, little right turns.

Haynes and his crew discuss power settings and headings for a few seconds.

Fitch: You lost the engine, huh?
Haynes: Yeah, well, yeah. It blew. We couldn't do anything about it. We shut it down.
Fitch: Yeah.
Flight Engineer: [*To SAM*] Go ahead with any help you can give us.
SAM: United Two thirty-two, understand that you have [the numbers] one and three engines operating. You have absolutely no hydraulic power. You have no control over the aircraft. Is that correct?
Haynes: [*To Fitch*] [I] can't think of anything that we [haven't] done. . . . There really isn't a procedure for this.
Fitch: No, the only thing I can think about that might help you at some point here [is to put] the [landing] gear down, and that might hold the nose down a bit.
SAM: Okay, United Two thirty-two, I've got Operational Engineering on its way over here, and at the present time you are doing just about everything that you can possibly do. Your flaps and slats, I believe, are in the up position, are they not?

APPROACH: When you get turned to that two-forty heading, sir, the airport will be about twelve o'clock and thirty-eight miles.
COPILOT: Okay, we're tryin' to control it just by power alone now. We have no hydraulics at all, so we're doing our best here.
APPROACH: Roger, and we've notified the [emergency] equipment out in that area, sir. The equipment is standing by.
FLIGHT ENGINEER: [*To SAM*] That is affirmative. That is affirmative. That is affirmative. Do you read?

Captain Haynes confirms certain frequencies and headings for the final approach.

HAYNES: Everybody ready?
FITCH: Anything above about two ten [knots] is going to give you a nose-up moment. . . .
HAYNES: We have almost no control of the airplane.
SAM: United Two thirty-two, in your handbook on page ninety-one, ninety-one . . .
COPILOT: We have no hydraulics at all.
HAYNES: It's gonna be tough, gonna be rough. . . .
FLIGHT ATTENDANT: So we're gonna evacuate?
HAYNES: Yeah. Well, we're gonna have the gear down.
FLIGHT ATTENDANT: Yeah.
HAYNES: And if we can keep the airplane on the ground and stop standing up, give us a second or two before you evacuate.
FLIGHT ENGINEER: [*To SAM*] We already have a no flap–no slat [approach] made up and we're gettin' ready. We're gonna try to put into Sioux City with gear down.
COPILOT: Okay . . . pull back a little more.
HAYNES: [*To flight attendant*] "Brace" will be the signal; it'll be over the PA system—brace, brace, brace.
SAM: United, you're tryin' to go into Sioux City. We'll contact Sioux City and have emergency equipment available.
FLIGHT ATTENDANT: [*To Haynes*] And that will be to evacuate?
HAYNES: No, that'll be to brace for landing.
FLIGHT ATTENDANT: Uh-huh.
HAYNES: And then if you have to evacuate you'll get the command signal to evacuate, but I really have my doubts you'll see us standing up, honey. Good luck, sweetheart.
FLIGHT ATTENDANT: Thank[s], you too.

FLIGHT ENGINEER: [*To SAM*] Okay, we will be tryin' to get in there [Sioux City].

SAM: Okay, United Two thirty-two, I'll stay with you.

FLIGHT ENGINEER: Okay, we will be waitin' in case you have anything more.

SAM: We're scurrying around, and I've got people out looking for more information.

HAYNES: The heading is two forty.

FITCH: Okay, I'm gonna try to hold you about two ten. I'll just see if it makes a difference if I bump it . . . bump it up in the air. This may be the world's greatest tricycle. . . .

FLIGHT ENGINEER: [*To crew after talking to flight attendant over interphone*] She says there appears to be some damage on that one wing. Do you want me to go back and take a look?

FITCH: No, we don't have time.

HAYNES: Okay, go ahead. Go ahead and see what you can see, not that it'll do any good. [*To Fitch*] I wish we had a little better control of the elevator. . . . They told us the autopilot would do this, but it sure as hell won't. Try yours again.

COPILOT: Can't get it on.

HAYNES: Well, we've got the ah . . . ah.

FITCH: All right, we came into the clear.

HAYNES: Turn, baby.

FITCH: Which way do you want it, Al?

HAYNES: Left.

COPILOT: Left.

HAYNES: Come on back, come on back, come on back. . . . As soon as that [is] vertical go for it, go for it. Watch that vertical speed the second it starts to move. Come back, come back, come back. Go for it. If we can get this under control elevator-wise we can work on steering later.

CABIN: [*Laughter*]

FITCH: We need to go left again to get ready to go. . . .

HAYNES: You keep goin' right two forty 'cause we still got two thousand feet to go. [*To Approach*] United Two thirty-two, we're gonna have to continue one more right turn. We got the elevators pretty much under control within three or four hundred feet, but we still can't do much with the steering.

APPROACH: United Two thirty-two heavy, roger. Understand you [have] the elevators possibly under control [enough to hold] altitude?

COPILOT: Barely.

HAYNES: Negative. We don't have it, but we are better, that's all.

APPROACH: Roger.

FITCH: You want to turn right?

HAYNES: Yeah, let's turn right.

FITCH: All right, here we go.

HAYNES: [*To Approach*] How far is the field now, please?

APPROACH: United Two thirty-two heavy, you're currently thirty-three miles northeast.

CABIN: [*Sound of three knocks on door*]

HAYNES: Thank you. [*To Fitch*] Just let her ease down. I wish they'd unlock that damn door [or] pull the circuit breaker on that door. Just unlock it, will ya?

CREW: Okay.

APPROACH: United Two thirty-two heavy, there are a couple of really small airports out in the vicinity here, and Storm Lake is four thousand two hundred feet by seventy-five [feet]. That's about fifteen miles east of your position.

FLIGHT ENGINEER: [*Comes back into the cockpit after his inspection of the back of the airplane*] All right, I walked to the back, and we got a lot of damage to the tail section. We could see through the window.

COPILOT: [*To Approach*] Roger, we're still goin' down [and] tryin' to control it. As we get down a little lower here we'll pick it out.

SAM: Okay, United Two thirty-two, you have a lot of damage to the tail section?

FLIGHT ENGINEER: The leading edge of the elevator is damaged. I mean, there's damage there that I can see. I don't know how much [there] is that I cannot see. I can see it on the leading edge, on the outer parts.

SAM: United Two thirty-two, Engineering is assembling right now, and they're listening to us.

Captain Haynes announces his orders to the passengers about the warning "Brace."

FLIGHT ENGINEER: [*To SAM*] Okay, [our] number-two engine blew. [There were] severe shudders and vibration through the airplane when it blew. Then we tried to pull the throttle back on number two, [but] it wouldn't come back. It was frozen. We shut it down

[and] turned off the fuel in that [engine and we] pulled the fire handle on it, and we have only been able to hold direction control through power application. . . . We're down to nine thousand [feet] now and we're tryin' to make Sioux City. We're gonna have to use alternate gear to get the [landing] gear down. I think we're gonna be kinda busy here. If there's anything I can talk to you about, I'll try to. If there's anything you can give for suggestions, give me a holler.

APPROACH: United Two thirty-two heavy, there is a small airport at twelve o'clock and seven miles. The runway is four thousand feet long there.

COPILOT: Hey, I'm controllin' it myself now. As soon as the captain gets back on, he'll give me a hand here. He's talking on the PA.

HAYNES: [*To the crew*] Okay, let's start this sucker down a little more.

FITCH: Okay, set your power a little bit.

HAYNES: Anybody have any ideas about [what to do about the landing gear]? He [the flight engineer] is talking to SAM.

FITCH: Yeah, he's talking to SAM. I'm gonna alternate-gear you. Maybe that will even help you. [But] if there is no [hydraulic] fluid, I don't know how the outboard ailerons are going to help you.

HAYNES: How do we get [landing] gear down?

FITCH: Well, they can free fall. The only thing is, we alternate the gear. We got the [landing gear] doors down?

HAYNES: Yep.

COPILOT: We're gonna have trouble stopping, too.

HAYNES: Oh, yeah. We don't have any brakes.

COPILOT: No brakes?

HAYNES: Well, we have some brakes [but not much].

FITCH: [Braking will be a] one-shot deal. Just mash it, mash it once. That's all you get. [*To Haynes*] I'm gonna turn ya. [I'm gonna] give you a left turn back to the airport. Is that okay?

HAYNES: I got it. [*To Approach*] Okay, United Two thirty-two, we're startin' to turn back to the airport. Since we have no hydraulics, braking [is] gonna really be a problem. [I] would suggest the [emergency] equipment be toward the far end of the runway. I think under the circumstances, regardless of the condition of the airplane when we stop, we're going to evacuate. So you might notify the ground crew [pretty much] that we're gonna do that.

APPROACH: United Two thirty-two heavy, wilco, sir. If you can continue that left turn to about two-twenty heading, sir, that'll take you right to the airport.

COPILOT: Two twenty, roger.

HAYNES: [*To Approach*] What's your ceiling right now?

FITCH: How far away are we from the airport? How far from the airport?

APPROACH: Ah, ceiling is four thousand, broken, and visibility's one five underneath it.

HAYNES: And the airport elevation?

APPROACH: One thousand ninety-eight [feet above sea level].

COPILOT: Well, [at] five thousand feet [of elevation] we ought to break out [of the clouds]. [*To Haynes*] If you have any problem about the spoilers, Al, we won't have those either, will we?

HAYNES: I don't think that'll help. I'm off for just a second to buckle up.

COPILOT: All right.

FITCH: You can tell me what you need. Holler what you need. . . .

HAYNES: What did SAM say? Good luck?

FLIGHT ENGINEER: He hasn't said anything.

HAYNES: Okay, well, forget them. Tell 'em you're leaving the air, and you're gonna come back up here and help us . . . and screw 'em. Ease her down just a little bit.

FITCH: When you get a chance ask them how far out we are?

HAYNES: [*To Approach*] How far are we away from the airport now . . . ?

APPROACH: Thirty-five miles, and if you continue that left turn about another fifteen or twenty degrees it'll take you right to the runway.

HAYNES: Okay. We don't have a localizer or a glide slope, so . . .

APPROACH: Yes, sir. You're well too far north of it now.

HAYNES: Okay.

SAM: United Two thirty-two, one more time. No hydraulic quantity, is that correct?

HAYNES: Now we gotta level off a little bit. We're six thousand feet above the field right [now]. About eighteen miles is where we want to be on the glide slope. We got about twelve miles to go before you want—

FLIGHT ENGINEER: [*To SAM*] Affirmative, affirmative, affirmative.

FITCH: Oh, yeah, we'll be with power.

FLIGHT ENGINEER: Add a couple of knots [of speed] for those speeds up there.

FITCH: All right.

HAYNES: Anybody got any idea about puttin' the [landing] gear down right now?

FITCH: All right, I would. I would suggest . . .

HAYNES: Should we free-fall it [the landing gear]?

COPILOT: Yeah, yeah. I got to get out of the way to get the door [to the manual release for the landing gear].

HAYNES: Put it down.

For the next few seconds the crew discusses the best way to get the landing gear down without hydraulic pressure. Either they will use gravity and let the gear fall out of the wells, or they will crank it down.

HAYNES: Okay, put it down.

FITCH: I don't know. I don't have any great ideas.

COPILOT: Try it out.

APPROACH: United Two thirty-two heavy, your present heading looks good.

HAYNES: [*To Approach*] We'll see how close we can come to holding it.

COPILOT: Apply a little power.

FITCH: I can slow you down. Do you want to go to one eighty-five [knots]?

HAYNES: Nope . . .

FLIGHT ENGINEER: [All] Green [on the landing gear down].

HAYNES: Go for it. Go ahead. Keep it at [one] eighty-five. Okay, start it down now. Ease back.

FLIGHT ENGINEER: Gear handle down. Gear handle down.

HAYNES: Okay, lock up and put everything away.

FITCH: Oh, shit, there go the slats.

HAYNES: A little right turn. Don't have much to do. Sit down and lock up. Get up there and see what he is doin' for power. . . . Okay, right turns. Level up first. Or level up your turn. Straighten out the turn. Get yourself all buttoned up.

CABIN: [*Sound of a groan*]

HAYNES: Level up here.

APPROACH: United Two thirty-two heavy, can you still make the slight right turns?

HAYNES: Yeah. Right turns are no problem, just left turns. . . .

APPROACH: Roger.

HAYNES: Well, mamma. We'll [make or miss] those baseball games after all.

COPILOT: Are you in good and tight?

FITCH: I'm not in at all, Bill.

COPILOT: No, not you. But him . . .

FITCH: Yeah. It seems [we're] gonna have to keep more power on the right engine.

APPROACH: United Two thirty-two heavy, sir, you are well too far north.

HAYNES: Pull it back. [*To Approach*] We know.

APPROACH: Two thirty-two heavy, your present heading is a little close, sir. Can you make a shallow left turn about ten degrees or so?

HAYNES: I'll try.

COPILOT: Back on the controls.

FITCH: Got to get my glasses on or I can't see shit.

COPILOT: [*To Approach*] Where' s the airport?

APPROACH: United Two thirty-two, the airport's currently twelve o'clock and two one [twenty-one] miles.

HAYNES: Twenty-one miles and thousand feet. We got to level off.

APPROACH: United Two thirty-two heavy, you're gonna have to widen out just slightly to your left, sir, to make the turn to final, and also it'll take you away from the city.

HAYNES: A little left bank. [*To Approach*] Whatever you do, keep us away from the city. [*To his crew*] Back, back.

FITCH: Hold this thing level if you can.

HAYNES: Level, baby, level, level . . .

FLIGHT ENGINEER: We're turning now.

FITCH: More power, more power, give 'em more power.

COPILOT: More power, full power.

FITCH: Power picks 'em up.

APPROACH: United Two thirty-two heavy, fly heading one eight zero, one eighty. . . .

HAYNES: I don't think we can do that, but we'll try. Can we turn left?

APPROACH: You are currently one seven miles northeast of the airport. You're doing good. . . .

HAYNES: It has to be a right turn to one eighty. We can't do anything about it. . . .

FLIGHT ENGINEER: Do you want this seat?

FITCH: Yes, do you mind?

FLIGHT ENGINEER: I don't mind. I think that you know what you were doing there before.

HAYNES: Level it off.

APPROACH: United Two thirty-two heavy, there's a tower five miles off to your right side that's three thousand four hundred . . . in height.

COPILOT: Roger.

FITCH: Keep turning right, Al. Keep turning right.

HAYNES: You got to level this sucker off. The only thing that I was afraid of was putting the gear down in case we have to ditch.

APPROACH: United Two thirty-two heavy, how steep a right turn can you make, sir?

HAYNES: About a thirty-degree bank.

APPROACH: United Two thirty-two heavy, roger. Turn right, heading one eight zero.

COPILOT: One eighty.

HAYNES: We got to level this sucker off. Come back, come back, come back.

FITCH: I got the tower [in view].

HAYNES: Come back, come back, all the way back.

FITCH: I can't handle that steep a bank. Can't handle that steep a bank.

APPROACH: United Two thirty-two heavy, [I've] been advised [that] there is a four-lane highway up in that area, sir, if you can pick that up.

HAYNES: Okay, we'll see what we can do here. We've already put down the gear, and we're gonna have to be puttin' [down] on something solid if we can. . . .

FITCH: Damn it. Wish we hadn't put that [landing] gear down.

FLIGHT ENGINEER: Ah, well.

FITCH: We don't know.

HAYNES: Just keep turnin' if you can.

FITCH: Which way do you want to go?

HAYNES: We got to go to one eighty. Right is the only way to go. So we can't control the airplane. That way . . . [*Laughs*]

COPILOT: All right, we're gonna have to try it straight ahead, Al. I think what we're gonna have to do—

APPROACH: United Two thirty-two heavy, if you can hold that altitude, sir, the right turn to one eighty would put you on about ten miles east of the airport.

HAYNES: That's what we're tryin' to do.

COPILOT: Let's see if we can get a shallow descent, Al.

HAYNES: That's what I'm tryin' to do. . . . Get this thing under control. When it starts up, push.

COPILOT: Okay. Here we go. Push hard, push hard.

FITCH: When the speed bleeds [creeps] back you'll catch it. Now, where do you want to go?

HAYNES: [I] want to keep turnin' right. Want to go to the airport.

FITCH: You want to go to the airport.

HAYNES: I want to get as close to the airport as we can.

FITCH: Okay.

HAYNES: If we have to set this thing down in dirt, we set it in the dirt. [*To copilot*] Get on the air and tell them we got about four minutes to go.

COPILOT: [*To Approach*] We've got about three or four minutes to go, [it] looks like.

HAYNES: PA system, PA system . . . Tell the passengers [to brace].

FLIGHT ENGINEER: [*On public-address system*] We have four minutes to touchdown, four minutes to touchdown. . . .

APPROACH: United Two thirty-two heavy, roger. Can you pick up a road or something up there?

COPILOT: We're tryin' it. [We're] still [fluctuating our altitude] anywhere from two thousand [feet] up to fifteen hundred feet down now, in waves.

FITCH: [*To Haynes*] Which way do you want to go?

HAYNES: Right, right, right. We gotta go—

FITCH: [The] airport's down there. Got it.

HAYNES: I don't see it yet.

COPILOT: Soon as it starts down, back we go. . . . Okay, now you can bring 'em up. There's the airport.

APPROACH: United Two thirty-two heavy, the airport is about eighteen miles southeast of your position, about two twenty on the heading, but we're gonna need you southbound away from the city first. If you can hold one-eighty heading . . .

COPILOT: [*To Approach*] We're tryin', tryin' to get to it right now.
FITCH: If I keep you about two hundred knots I seem to be able to get enough control. . . .
APPROACH: United Two thirty-two heavy, advise if you can pick up a road or anything where you can possibly land it on that.
HAYNES: [*To Approach*] Okay, we're a hundred-eighty-degree heading. Now what do you want?
APPROACH: United Two thirty-two, if you can hold the altitude, the one-eighty heading will work fine for about seven miles.
HAYNES: Okay, we're tryin' to turn back. [*To Fitch*] Forward, make a left turn, left. . . .
FITCH: No left at all.
HAYNES: No left at all?
FITCH: I'll give you some.
COPILOT: Okay, now it's—
HAYNES: Back, back, back, back . . . forward, forward, forward . . . Won't this be a fun landing? Back . . . [*Laughs*]
APPROACH: United Two thirty-two heavy, can you hold that heading, sir?
COPILOT: Yeah, we're on it now for a little while.
APPROACH: United Two thirty-two heavy, roger. That heading will put you currently fifteen miles northeast of the airport. If you can hold that, it'll put you on about a three-mile final.
COPILOT: Okay, we're givin' it heck.
HAYNES: I'll tell you what. I'll write off your damn PC [Pilot Certificate] if we make this . . . *when* we make this. Hold the heading if you can. . . . That's fine. Turn left. Help me turn left so we know what it's doing. Back, back, back . . .
APPROACH: United Two thirty-two heavy, the airport's currently twelve o'clock and one three [thirteen] miles.
COPILOT: Okay, we're lookin' for it.
HAYNES: Forward, forward, forward . . .
COPILOT: Twelve o'clock at thirteen miles. We have to start down, but . . .
FITCH: Ask what the field elevation is.
HAYNES: [*To Approach*] Field elevation is what again?
COPILOT: A thousand eighty.
APPROACH: Ah, eleven hundred feet, one thousand one hundred . . .
HAYNES: Okay, thank you.
FITCH: Let's start down. We have to ease it down.

HAYNES: [*To Approach*] We're startin' down a little bit now. We got a little better control of the elevator.

APPROACH: United Two thirty-two heavy, roger. The airport's currently at your one o'clock position, one zero [ten] miles.

HAYNES: Ease it down, ease it down. . . .

CABIN: [*Sound of groan; sound of exhalation*]

FITCH: I got the runway, if you don't. . . .

HAYNES: I don't. . . . Come back, come back.

FITCH: It's off to the right over there.

COPILOT: Right there. Let's see if we can hold [a descent rate of] five hundred feet a minute.

APPROACH: United Two thirty-two heavy, if you can't make the airport, sir, there is an interstate that runs north to south, to the east side of the airport. It's a four-lane interstate.

FITCH: See? We got [the] Tower [in sight] right here at our one o'clock low. . . .

HAYNES: [*To Approach*] We're just passing it [the highway] right now. We're gonna try for the air[port]. [*To Fitch*] Is that the runway right there? [*To Approach*] We have the runway in sight. We have the runway in sight. We have the runway in sight. We'll be with you shortly. Thanks a lot for your help.

FITCH: Bring it on down. . . . Ease 'er down.

COPILOT: Oh, baby.

FITCH: Ease her down.

HAYNES: Tell 'em [the passengers] that we're just two minutes from landing.

APPROACH: United Two thirty-two heavy, the wind's currently three six zero at one one three sixty at eleven. You're cleared to land on any runway. . . .

HAYNES: [*Laughs*] Roger. [*Laughs*] You want to be particular and make it a runway, huh?

FLIGHT ENGINEER: [*On public-address system to passengers*] Two minutes.

In the passenger cabin, the flight attendants begin shouting for passengers to get their heads down. The first flight attendant yells commands.

FITCH: What's the wind?

HAYNES: [*To Approach*] Say the wind one more time.

APPROACH: Wind's zero one zero at one one. . . .
COPILOT: Yeah, we want to go down.
FITCH: Yeah, I can see the runway, but I got to keep control on ya.
COPILOT: Pull it off a little.
HAYNES: See if you can get us a left turn.
COPILOT: Left turn just a hair, Al.
HAYNES: [*To Approach*] Okay, we're all three talking at once. Say it [the wind] again one more time.
APPROACH: Zero one zero at one one, and there is a runway that's closed, sir, that could probably work to the south. It runs northeast to southwest.
HAYNES: We're pretty well lined up on this one here. . . .
FITCH: I'll pull the spoilers [speed brakes] on the touch[down].
HAYNES: Get the brakes on with me.
APPROACH: United Two thirty-two heavy, roger, sir. That's a closed runway, sir, that'll work, sir. We're gettin' the equipment off the runway. They'll line up for that one.
HAYNES: How long is it?
APPROACH: Sixty-six hundred feet, six thousand six hundred. Equipment's comin' off.
HAYNES: [*To crew*] Pull the power back. That's right. Pull the left one [throttle] back.
COPILOT: Pull the left one back.
APPROACH: At the end of the runway it's just wide-open field.
COCKPIT UNIDENTIFIED VOICE: Left throttle, left, left, left, left . . .
COCKPIT UNIDENTIFIED VOICE: God!
CABIN: [*Sound of impact*]

END OF TAPE

The first flight attendant would later describe the impact in the passenger cabin as "a wind coming over a field and blowing down the grass." All the passengers' heads were down. A couple of passengers put up their heads prior to impact, and the flight attendants yelled at them to keep their heads down. There was an "unbelievable noise" during the impact, according to the first flight attendant. She saw two overhead bins begin to open, and she felt a fireball pass through the cabin. The fireball came up the left side of the cabin and went three quarters of the way over the first flight attendant, and a few seconds later the airplane came to a stop. She was upside down and pressing against her seat belt. She released her seat belt with difficulty. She slid down and stood up on the debris of the airplane, but she was able to stand upright. She saw a man lying forward of where she was standing, and his ankles were wedged in a crevice of the wreckage. She pulled him free. She saw a crack of light and heard passengers say there was an opening. She went to the opening and started to push passengers out. She thought the entire area of the airplane forward of her jumpseat in the first-class cabin was gone. There were "no doors . . . nothing recognizable." She saw gray-black smoke coming from the mid-cabin. It was "swirling, rolling forward." She exited the airplane into a cornfield and ran from the airplane. She tried to find a clear area where she could gather the survivors, but because of the height of the cornstalks, she could not see a clear area. She started tramping down the cornstalks. She saw another flight attendant; he seemed dazed.

Flight attendant number two remembered the airplane hitting the ground this way: "Kind of screeching back and forth and back and forth and the lights going out and coming back on." There was a lot of "skidding around." She did not remember anything after that. She was not buckled in at impact. She was found with dirt in her mouth and nose in an area where other people were not alive. She was seriously injured.

Flight 232 touched down on the threshold and to the left of the center line on closed Runway 22. The first contact with the ground was made by the right wing tip. The right main landing gear broke the concrete beyond the first impact point. The airplane skidded off Runway 22, then flipped over. The main fuselage section stopped upside down in a cornfield between Runway 17 and Taxiway L. Both wings were still attached, though most of the right wing had broken off. The tail section had broken off at the first impact and continued to slide and tumble down the runway and came to a stop on Taxiway L.

The main fuselage section and the tail section came to a stop 3,650 feet from the point of first impact.

One passenger, a thirty-six-year-old man who assumed the brace position prior to impact, later described it as a little harder than a normal landing. But the second impact was "so hard that I thought I would go through the seat cushion." There was a "tremendous" explosion, and he was thrown around in his seat. The lights went out. The floor emergency lights flickered, then went off. He saw smoke and there was a "terrible smell." He was struck by other passengers who fell from their seats. Suitcases tumbled from the overhead bins. He got out of his seat and walked on the ceiling through an opening in the fuselage.

Just before impact, a twenty-five-year-old mother placed her twenty-three-month-old son on the floor and covered him with a blanket. She wedged a pillow between his head and the cabin wall, then took the brace position and held her son down with one hand on his legs and her other hand on his chest. The plane hit the ground and rolled to the right. She heard metal scraping, and dirt and plastic flew around her. "My son flew up in the air, and I managed to grab ahold of him around the waist," she said later. "He struck his head several times before the plane came to a stop, and several times I had to pull him back into my arms as he slid out of my grip." When the plane stopped, they were upside down, and she was holding on to her son. When she released her seat belt, they fell to the ceiling of the upside-down cabin. She landed on her shoulder, and her son hit his head. A man seated next to her helped her up and pushed her toward the hole in the fuselage that other passengers were using to evacuate the wreckage. She walked barefoot through plastic and luggage. "As I exited," she said, "I looked to my right and saw an elderly gray-hair [*sic*] woman still buckled in her seat. I pointed her out to the person behind me, but I saw no movement from the lady."

There were 296 passengers and crew aboard Flight 232. One hundred eighty-five people survived the accident, including 7 flight attendants and the 3 cockpit members. Fatalities included 110 passengers and 1 flight attendant.

GLOSSARY

ACARS (Air-to-ground Communication and Reporting System): A cockpit messaging system from ground to aircraft that can be read as a display or can be printed out on a cockpit printer. Among other messages received on ACARS are the connecting gate numbers, which the cabin crew often reads out to passengers before landing.

ANNUNCIATOR LIGHT: A caution or warning light on the panel of the cockpit called the annunciator panel. There is an annunciator light for every system on the aircraft.

APU (Auxiliary Power Unit): An engine aboard the aircraft, usually located in the tail, for the generation of electrical and hydraulic power when the main engines are shut off.

ARMED: Set. "Spoilers armed," for instance, indicates that the spoilers have been programmed to deploy automatically without pilot involvement after landing when the main wheels are spinning at a certain rate of speed.

AUTOPILOT: A device that takes raw data and gives the aircraft a perfect flight path. The autopilot takes directions from the flight director computer system, which commands the aircraft from takeoff to landing.

BEFORE LANDING CHECKLIST: This means that landing gear is down and there is a three-light verification of such. The spoiler handle is armed, and when the aircraft touches down, the spoilers deploy—i.e., they elevate on the wings to kill lift—when the main wheels reach a certain set speed. When the spoilers are deployed while the aircraft is still in flight, they are referred to as speed brakes and function the same way, by spoiling lift and slowing down the aircraft.

BLEEDS: Use of heated air in a jet engine to pressurize and heat the cabin and operate pneumatic systems. When bleeds in the engine are open, air is supplied to these systems.

BUG: A setting on the rims of several instruments in the cockpit to remind the pilot of reference speeds and altitudes. Bug speed, for instance, is the aircraft's determined approach speed based on the weight of the airplane. There are other bug speeds, but approach speed is the most important one.

CHARLIE, CHARLIE: Yes, yes.

COPY: "I understand," usually a remark made by the cockpit crew while communicating with flight controllers on the ground.

CROSS FEEDS: Valves that permit the transfer of fuel from one tank to another.

DME (Distance-Measuring Equipment): This provides a display in nautical miles of distance from a VOR or an ILS (Instrument Landing System) beam at an airport.

FLAPS: Trailing-edge devices on the wings that increase the lift of the wings on takeoff and especially on landing.

FLIGHT DIRECTOR: A computer system that flies the airplane automatically and corrects for wind and drift and flies a course or beam, laterally and vertically. It is programmed at the start of a flight to operate the airplane through to the end of the flight; with this responsibility, it directs the airplane's autopilot.

GLIDE SLOPE CAPTURE: At the proper altitude and on a correct heading for a safe landing on an Instrument Landing System (ILS) approach to a runway.

HEAVY: An air-traffic controller's term for a larger aircraft; usually a wide-bodied Boeing 747, DC-10, or L-1011, but also in some instances a reference to a stretch DC-8, with a takeoff weight in excess of 25,000 pounds.

HYDRAULICS: The mechanism (hydraulic fluid under pressure) that controls movable surfaces (elevators, rudders, ailerons, flaps, etc.) on the aircraft, as in "Do we have hydraulics?"

LNAV (Lateral Navigation): Modern airplanes, like a Boeing 767, have LNAV and VNAV functions. LNAV navigates an aircraft automatically in a lateral direction. VNAV will navigate the aircraft automatically in a vertical direction. Both LNAV and VNAV are computer generated.

LOCALIZER: An electronic beam that indicates the runway center line; part of the Instrument Landing System (ILS).

MINIMUMS: Levels of altitude and visibility below which the aircraft cannot safely fly.

OUTER MARKER: A point of instrument light in the cockpit and a sound warning to indicate the final approach, telling pilots they have begun the serious part of the approach, or glide slope capture; part of the Instrument Landing System (ILS). Next is the middle marker, reached at about 200 feet above the ground. An inner marker exists for some approaches.

POSITIVE (CLIMB RATE): On takeoff, before the landing gear is raised, there must be an indication of a positive climb rate by a vertical velocity indicator that shows a climb and an altimeter that shows a climb in feet.

PUMPS ARE UP: Hydraulic pumps are on high. Jets take off and land with pumps at a high system pressure to raise the heavy landing gear, and above 18,000 feet the pumps are set at a lower position for cruise.

ROTATE: To lift the aircraft's nose off the runway during takeoff when the required speed is reached, depending on the aircraft's weight.

RUDDER RATIO: Limits the amount of rudder force that can be used at high speed. The ratio is smaller at high altitudes and greater at lower altitudes, especially when the aircraft is landing.

SINK RATE: A rate of descent.

SLATS: Leading-edge devices on the wings that give the wings additional lift and act in concert usually with the trailing-edge wing flaps.

SPOILERS: Surfaces on the tops of the wings that are extended after touchdown to kill lift on the wings. Spoilers are deployed sometimes in flight to "brake" the airplane and reduce altitude, and in this use they are referred to as speed brakes.

SPOOL UP: Increase the rpm of the engines.

STALL SPEED: Airspeed at which the aircraft's wings can no longer maintain lift to prevent the aircraft from falling. To recover from a stall, the aircraft must have sufficient altitude to regain adequate airspeed by putting the nose down and adding power.

STICK SHAKER: A warning device on commercial aircraft that literally shakes the control column in the pilot's hands to warn of the approach of the aircraft's stall speed.

THREE GREEN: Means that all three sets of landing gear are down and locked.

TRANSPONDER: An electronic device aboard aircraft that communicates a signal to air-traffic controllers' radar and other aircraft indicating an identifying number and usually the aircraft's altitude and sometimes its direction.

VECTORS: Compass headings issued by ground controllers.

VEE ONE: A decision speed on takeoff at which the pilot can either stop the aircraft on the runway if there is an engine failure or take off and clear the end of the runway by thirty-five feet.

VEE TWO: A speed at which the aircraft can continue to climb on one engine in case of an engine failure during this critical part of flight.

VEE ROTATE: Point at which the aircraft has reached sufficient speed to pull the nose off the runway.

VISUAL ADVISORY WARNING: Another term for an annunciator light that goes on to warn of a system failure or malfunction.

VOR (Very high frequency Omnidirectional Radio range): An old navigational aid: a radio beacon described as the radial spokes in a 360-degree wheel.

WAKE VORTEX: Vortex created by the lift of a wing. A dangerous time is when a plane is "slow and clean," when the flaps are still up and it is ready for landing; wings are working their hardest. This creates a violent, spiraling wind, called a vortex, like a little tornado. It is vicious, and a larger airplane can upset a smaller airplane with a wake vortex.

YAW DAMPER: A mechanical device on the control surfaces in the tail of the aircraft that helps alleviate the tendency of an aircraft's nose to yaw left or right, which in turn can induce a "Dutch roll."